11-14-95

Geometric Concepts
for Geometric Design

Advisory Board

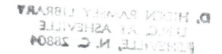
Geometric Concepts
for Geometric Design

Wolfgang Boehm
Technische Universität Braunschweig
Braunschweig, Germany

Hartmut Prautzsch
Universität Karlsruhe
Karlsruhe, Germany

A K Peters
Wellesley, Massachusetts

Editorial, Sales, and Customer Service Office

A K Peters, Ltd.
289 Linden Street
Wellesley, MA 02181

Library of Congress Cataloging-in-Publication Data

Boehm, Wolfgang, 1928-
     Geometric Concepts for Geometric Design / Wolfgang Boehm, Hartmut Prautzsch.
         p. cm.
     Includes bibliographic references and index.
     ISBN 1-56881-004-0
     1. Geometry. 2. Geometry—Data processing. I. Prautzsch, Hartmut. II. Title.
     QA445.B63   1993b
     516—dc20   93-20666
                                                                     CIP

About the cover: The Cover picture shows a computer generated shaded image of a chalice based on a drawing by Paolo Ucello (1397-1475). Ucello's hand drawing was the first extant complex geometrical form rendered according to the laws of perspective. (Perspective Study of a Chalice, Drawing, Gabinetto dei Disegni, Florence, ca 1430-1440.)

Printed in the United States of America
97 96 95 94 93 10 9 8 7 6 5 4 3 2 1

Contents

IV Euclidean Geometry {#iv-euclidean-geometry} 153

VI Some Descriptive Geometry   265

24 Associated Projections

25 Penetrations

VII Basic Algebraic Geometry   297

26 Implicit Curves and Surfaces

VIII Differential Geometry   351

Preface

This book addresses students, teachers and researchers in mathematics, computer science and engineering who are confronted with geometric problems, attracted by their beauty, and/or wish to get a deeper geometric background.

Its purpose is to give a solid foundation of geometric methods and their underlying principles. It may serve as an introduction to geometry as well as a practical guide to geometric design and modeling and to other applications of geometry.

The main idea of this book is to provide an imagination for what happens geometrically and to present tools for describing problems. A problem is often solved simply by finding the right description. The topics presented have been chosen from the many geometric problems the authors have confronted during their work in applied geometry and geometric design.

In writing this book we intended to disconnect geometric ideas and methods from special applications, in order to make these ideas clear and to allow the reader to apply the presented material to other problems of a geometric nature. Also, in many situations, a figure can say more than a thousand words. This old Chinese proverb ought to be a guideline in writing a text on geometry. Therefore, figures are crucial throughout this book, while diagrams are an integral part of Chapters 1 and 28.

This book owes its inception to lectures given by Boehm at Rensselaer Polytechnic Institute and the Technical University of Braunschweig several times between 1986 and 1990. This book has been partly written at Rensselaer, and we are greatly indebted to Harry McLaughlin who has been promoting Applied Geometry at Rensselaer and who together with Joe Ecker initiated their cooperation with the TU Braunschweig. Andreas Johannsen read the first and later drafts of the book very carefully, and we benefitted much from his helpful suggestions. We thank Dr. Michael Kaps and Wolfgang Völker for typing the manuscript; Daniel Bister for proof reading the mathematics and Jeannette Machnis for proofreading the English text; and Mrs. Diane McNulty for her judicial and committed assistance in the cooperation with Rensselaer.

Troy, in December 1992

Wolfgang Boehm

Hartmut Prautzsch

Notation

The following notation is used throughout this book:

Scalars $\qquad\qquad\qquad\qquad\qquad\qquad\quad$ $\alpha, \beta, \ldots, a, b, \ldots$

Vectors, points, coordinate columns $\qquad\quad$ $\mathbf{a}, \mathbf{b}, \ldots, \mathbf{p}, \mathbf{q}, \ldots$

Extended columns (by an additional coordinate) $\qquad$ $\mathsf{x}, \mathsf{y}, \ldots$

Differences between two points $\qquad\qquad$ $\triangle\mathbf{x}, \triangle\mathbf{y}, \ldots, \triangle\mathsf{x}, \triangle\mathsf{y} \ldots$

Matrices $\qquad\qquad\qquad\qquad\qquad\qquad$ $A, B, \ldots$

Augmented matrices $\qquad\qquad\qquad\qquad$ $\mathbb{A}, \mathbb{B}, \ldots$

Vector spaces $\qquad\qquad\qquad\qquad\qquad$ $\mathbf{V}, \mathbf{A}, \ldots$

Point spaces $\qquad\qquad\qquad\qquad\qquad\quad$ $\mathcal{A}, \mathcal{P}, \ldots$

Orthogonal angles $\qquad\qquad\qquad\qquad$ ⌐

Parallelism $\qquad\qquad\qquad\qquad\qquad\quad$ $\parallel$

Bold type is used whenever a new term is introduced.

Remark: Each chapter starts with an abstract and a short bibliography for further information on the particular subject. The complete references are listed at the end of the book.

PART ONE

Some Linear Algebra

Many problems encountered in applied mathematics are linear or can be approximated by linear systems which are, in general, computationally tractable. The corresponding mathematical subdiscipline is called **linear algebra**. At the heart of linear algebra are techniques, such as Gaussian elimination and the Gauss-Jordan algorithm, for computing solutions of linear systems. The main tool of linear algebra is matrices which help to arrange coefficients and describe operations.

1 Linear Systems

Most finite linear systems can be described by matrices, a very useful short-hand notation which emphasizes the underlying linear structure and the interdependencies between the equations.

Literature: Atkinson, Boehm·Prautzsch, Conte·de Boor

1.1 Matrix Notation

A **linear system** is a set of equations of the form

$$a_{1,1}\,x_1 + \cdots + a_{1,n}\,x_n = a_1$$

$$\vdots \qquad\qquad \vdots \quad \vdots$$

$$a_{m,1}\,x_1 + \cdots + a_{m,n}\,x_n = a_m \ ,$$

where the a's are given real numbers and the x's are unknowns. The array A of the coefficients $a_{i,k}$,

$$A = \begin{bmatrix} a_{1,1} & a_{1,2} & \cdots & a_{1,n} \\ a_{2,1} & a_{2,2} & \cdots & a_{2,n} \\ \vdots & \vdots & \ddots & \vdots \\ a_{m,1} & a_{m,2} & \cdots & a_{m,n} \end{bmatrix} = [a_{i,k}] \ ,$$

is called an $m \times n$ **matrix**. The matrix A contains the element $a_{i,k}$ in its ith row and kth column. Similarly, the a_i can be written as an $m \times 1$ matrix or m **column**,

$$\mathbf{a} = \begin{bmatrix} a_1 \\ a_2 \\ \vdots \\ a_m \end{bmatrix} .$$

Consequently one has

$$A = [\mathbf{a}_1 \ \ldots \ \mathbf{a}_n] \ ,$$

where $\mathbf{a}_k$ represents the kth column of A. Note that a **scalar** a can be viewed as a 1×1 matrix.

The $n \times m$ matrix $A^t = \left[a_{i,k}^t \right]$, defined by $a_{i,k}^t = a_{k,i}$, is called the **transpose** of A, e.g., for A above

$$A^t = \begin{bmatrix} a_{1,1} & \cdots & a_{m,1} \\ \vdots & \ddots & \vdots \\ a_{1,n} & \cdots & a_{m,n} \end{bmatrix} = [a_{k,i}] \ .$$

In particular, the transpose $\mathbf{a}^t$ of an m column $\mathbf{a}$ forms an m **row**

$$\mathbf{a}^t = [a_1 \ \ldots \ a_m] \ .$$

Often it is helpful to visualize an $m \times n$ matrix A or an m column $\mathbf{a}$ in **block form**, i.e., as a rectangle of height m and width n or 1, respectively:

The matrix A is a **square matrix** if $m = n$, and it is **symmetric** if additionally $a_{i,k} = a_{k,i}$. A square matrix $[u_{i,k}]$ is called **upper triangular** if $u_{i,k} = 0$ for $i > k$. Similarly a square matrix $[l_{i,k}]$ is called **lower triangular** if $l_{i,k} = 0$ for $i < k$. The **Kronecker symbol**

$$\delta_{i,k} = \begin{cases} 1 & \text{if } i = k \\ 0 & \text{otherwise} \end{cases}$$

is used to define the **identity matrix** as the $n \times n$ square matrix $I = [\delta_{i,k}]$.

1.2 Matrix Multiplication

Let $A = [a_{i,j}]$ be an $m \times l$ matrix and $B = [b_{j,k}]$ an $l \times n$ matrix. The $m \times n$ matrix $C = [c_{i,k}]$ with the entries

$$c_{i,k} = \sum_{j=1}^{l} a_{i,j}\, b_{j,k}$$

is called the **product** AB of A and B, in this order. Note that the width l of A has to match the height l of B. It is helpful to visualize the product $AB = C$ in block form, as introduced above:

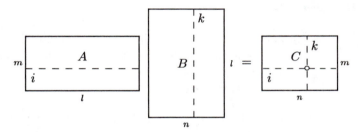

The element $c_{i,k}$ is the **dot** or **scalar product** of the ith row of A with the kth column of B. This may be memorized as "**row times column**".
Using this product the linear system in Section 1.1 can be written more compactly as

$$A\mathbf{x} = \mathbf{a} ,$$

and visualized by blocks as

where $\mathbf{x}$ denotes the n column of the unknown x_i. Likewise the **scalar product** α of two m columns $\mathbf{a}$ and $\mathbf{b}$ can be written as

$$\alpha = \mathbf{a}^{t}\mathbf{b} = \mathbf{b}^{t}\mathbf{a} .$$

Note that the product $\mathbf{x}\alpha$ is defined as a matrix multiplication, but $A\alpha$ is not. It is convenient to define $A\alpha = \alpha A$ as the matrix of elements $a_{i,k}\alpha$, i.e., one has

$$A \cdot \alpha = [\mathbf{a}_1\alpha \ldots \mathbf{a}_n\alpha] = [a_{i,k}\alpha] \ .$$

In particular, one gets for $\alpha = 0$ the **null column** $\mathbf{o} = \mathbf{x}0$ and the **null matrix** $O = 0\,A$.

A square matrix A is said to be **non-singular** if its **inverse** A^{-1} defined by $A^{-1}A = AA^{-1} = I$ exists. Finally, a matrix B is said to be **orthonormal** if $B^tB = I$.

1.3 Gaussian Elimination

Linear systems are most frequently solved by **Gaussian elimination**. It is convenient to represent the linear system $A\mathbf{x} = \mathbf{a}$ by the **augmented matrix**

$$[A|\mathbf{a}] = [\mathbf{a}_1 \ldots \mathbf{a}_n\,|\mathbf{a}] \ .$$

Then the linear systems obtained by the following simple operations on $[A|\mathbf{a}]$ will have the same solutions:

    1 exchanging two rows,
    2 multiplying one row by a factor $\neq 0$,
    3 adding one row to another,
    4 exchanging two columns of A while simultaneously
        exchanging the corresponding unknowns in the column $\mathbf{x}$.

It was Gauss' idea to use these four simple operations to transform $[A|\mathbf{a}]$ into the matrix $[B|\mathbf{b}]$, where B is composed of an upper triangular, non-singular $r \times r$ matrix U, an $r \times n{-}r$ matrix matrix B^*, an $m{-}r \times n$ null matrix, an r column $\mathbf{b}$, and an $m{-}r$ column $\mathbf{s}$, as shown below.

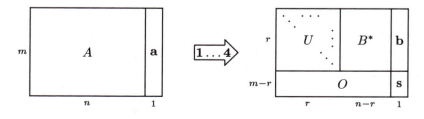

If $\mathbf{s} \neq \mathbf{o}$, there exists no solution. However, if $\mathbf{s} = \mathbf{o}$, there exists an $n-r$ parameter family of solutions which can easily be determined from the equivalent system $B\mathbf{x} = \mathbf{b}$ as follows. Assigning arbitrary values to $x_{r+1}, \ldots, x_n$ as parameters one can compute x_r **backward** from row r, then x_{r-1} from row $r-1$, ..., and finally x_1 from row 1.

Remark 1: The number r is called the **rank** of A, denoted by $rank\,A$. Note that $r \leq m$ and $r \leq n$.

Remark 2: If $\mathbf{a} = \mathbf{o}$, the linear system $A\mathbf{x} = \mathbf{a}$ is called **homogeneous**. Then one also has $\mathbf{b} = \mathbf{o}$. The homogeneous system has a non–trivial solution if and only if $r < n$ as can be inferred from the equivalent system $B\mathbf{x} = O$. If $\mathbf{x}$ is a solution of a homogeneous system, then $\mathbf{x} \cdot \varrho$, where $\varrho \neq 0$, is also a solution.

1.4 Gauss-Jordan Algorithm

Gaussian elimination can further be used to construct an explicit representation for the set of all solutions of the linear system $A\mathbf{x} = \mathbf{a}$. With the aid of the operations **1**, **2**, **3**, one transforms the matrix $[U\,|\,B^*\,|\,\mathbf{b}]$ from above into $[I\,|\,C^*\,|\,\mathbf{c}^*]$ as illustrated in the following diagram.

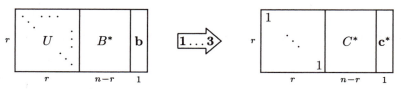

The general solution of this system is depicted below where $-I$ denotes the negative $n{-}r \times n{-}r$ identity matrix and $\mathbf{t}$ an $n - r$ parameter column.

Note that for $r < n$ the representation of $\mathbf{x}$ depends on the sequence of operations performed during Gaussian elimination.

Remark 3: The construction can be reversed. Let $[C|\mathbf{c}]$ represent the set

$$\mathbf{x} = \mathbf{c} + C\mathbf{t}$$

of (given) solutions where $\mathbf{c}$ is some n column, C an $n \times m$ matrix, and $\mathbf{t}$ a column of m free parameters. The set represented by $[C|\mathbf{c}]$ does not change if the transposed matrix $[C|\mathbf{c}]^t$ is modified by Gaussian elimination. Using the operations $\mathbf{1}, \ldots, \mathbf{4}$, the matrix C^t can be transformed into an $s \times n$ matrix $[-I|D^t]$ provided *rank* $C = s$. This is illustrated below where the superfluous zero rows are discarded. Adding appropriate multiples of rows of $[-I|D^t]$ to $\mathbf{c}^t$ one obtains a row $\begin{bmatrix} \mathbf{o}^t & \mathbf{d}^t \end{bmatrix}$ as illustrated below.

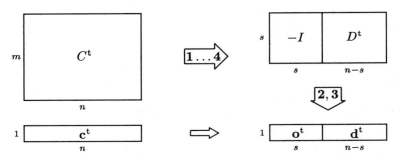

Now one easily obtains a linear system for which $\mathbf{x} = \mathbf{c} + C\mathbf{t}$ is a solution, namely

1.5 LU-Factorization

Often the matrix A of a linear system is square, i.e., $m = n$. Such a system is uniquely solvable if and only if A is non-singular or equivalently if $rank\ A = n$.

A non-singular matrix A can sometimes be factored into a lower-triangular matrix L, whose diagonal entries are all equal to 1, and an upper triangular matrix U, i.e., $A = LU$,

$$
\begin{bmatrix} a_{1,1} & \cdots & a_{1,n} \\ \vdots & A & \vdots \\ a_{n,1} & \cdots & a_{n,n} \end{bmatrix}
=
\begin{bmatrix} 1 & & \\ \vdots & L & \\ l_{n,1} & \cdots & 1 \end{bmatrix}
\begin{bmatrix} u_{1,1} & \cdots & u_{1,n} \\ & U & \vdots \\ & & u_{n,n} \end{bmatrix}
$$

The entries of L and U can successively be computed by means of the matrix multiplication rule for $a_{1,1}, \ldots, a_{1,n}, a_{2,1}, \ldots, a_{2,n}, \ldots, a_{n,1}, \ldots, a_{n,n}$ in this order. At each step there is exactly one unknown $l_{i,k}$ or $u_{i,k}$ to be determined.

If a non-singular matrix A cannot be factored in this way, one can always rearrange the rows of A to obtain a matrix A^* which has an **LU-factorization**. In all cases one can start to compute L and U as if A were to be factored and interchange the rows of A during the computation whenever it becomes necessary to avoid dividing by zero. The LU-factorization is another organization of Gaussian elimination and can be used to solve a system $Ax = a$. Let $[A^*\,|\,a^*]$ be obtained from $[A\,|\,a]$ by a row permutation such that an LU-factorization $A^* = LU$ exists. Solving the two triangular systems

$$
\begin{bmatrix} 1 & & \\ \vdots & L & \\ l_{n,1} & \cdots & 1 \end{bmatrix}
\begin{bmatrix} b_1 \\ \vdots \\ b_n \end{bmatrix}
=
\begin{bmatrix} a_1^* \\ \vdots \\ a_n^* \end{bmatrix}
\quad \text{and} \quad
\begin{bmatrix} u_{1,1} & \cdots & u_{1,n} \\ & U & \vdots \\ & & u_{n,n} \end{bmatrix}
\begin{bmatrix} x_1 \\ \vdots \\ x_n \end{bmatrix}
=
\begin{bmatrix} b_1 \\ \vdots \\ b_n \end{bmatrix}
$$

by **forward** and **backward substitution** respectively, yields the solution for $A^* x = a^*$ and hence for $Ax = a$.

Remark 4: The LU-factorization is useful for solving the system $A\mathbf{x} = \mathbf{a}$ repeatedly for a fixed coefficient matrix A and different right hand sides $\mathbf{a}$. In particular, if the right hand sides are the columns of the identity matrix I one obtains the inverse of A.

Remark 5: If A is symmetric and $\mathbf{x}^t A\mathbf{x} > 0$ for all $\mathbf{x} \neq \mathbf{0}$, then A is called positive definite, and a symmetric factorization $A = C^t C$, where C is an upper triangular matrix, is possible without row interchanges. This is called a **Cholesky factorization.**

1.6 Cramer's Rule

Let $A = [a_{i,k}]$ be a square $n \times n$ matrix and $A_{i,k}$ the submatrix obtained from A by deleting the ith row and kth column. Then the **determinant** of A, written $\det A$, is defined by the recursion

$$\det A = \sum_{k=1}^{n} (-1)^{i+k} a_{i,k} \det A_{i,k} \quad \text{and} \quad \det [a] = a,$$

for any scalar a. This definition does not depend on the choice of i and is called **Laplace expansion** along the ith row. The term $(-1)^{i+k} \det A_{i,k}$ is called the **cofactor** of $a_{i,k}$.

The determinant can be used to solve a non-homogeneous linear system $A\mathbf{x} = \mathbf{a}$ when A is some non-singular $n \times n$ matrix .

Let $A_k = [\mathbf{a}_1 \ldots \mathbf{a} \ldots \mathbf{a}_n]$ be obtained from $A = [\mathbf{a}_1 \ldots \mathbf{a}_k \ldots \mathbf{a}_n]$ by replacing the kth column with $\mathbf{a}$. Then **Cramer's rule,**

$$x_k = \frac{\det A_k}{\det A}, \quad k = 1, \ldots, n ,$$

gives the coordinates x_k of the solution. Note that $\det A \neq 0$ whenever A is non-singular.

In the case of a homogeneous system $A\mathbf{x} = \mathbf{o}$, where A is an $n-1 \times n$ matrix with $rank\, A = n - 1$, one can show that

$$x_k = \varrho \cdot (-1)^k \det A_k^*, \quad k = 1, \ldots, n ,$$

provides the solution of the system, where $\varrho \neq 0$ is a free parameter and A_k^* is obtained from A by deleting the kth column.

Remark 6: Cramer's rule is of practical use only for small n.

1.7 Notes and Problems

1 It is possible to improve the numerical stability of Gaussian elimination by row interchanges.

2 One of the numerically most stable algorithms used to solve linear systems is the so-called **Householder algorithm**.

3 Let A^* and I^* be obtained from the matrices A and I of equal height by the same row permutation. Then $P = I^*$ can be used to write down the permutation of A, i.e.,

$$A^* = PA \ .$$

P is called a permutation matrix. It is inverse to its transpose, i.e., $P^{\mathrm{t}}P = I$.

4 Using a permutation matrix P (see Note 3), Gaussian elimination can be summarized as

$$L^{-1}P[A\,|\,\mathbf{a}] = U \ .$$

5 The LU-factorization of $A^* = PA$ can be used to compute $X = A^{-1}$ by solving $A^*X = P$ column by column.

6 Most elimination methods for solving linear systems are actually just different organizations of the Gaussian elimination process. They differ only in the ordering of the computation steps.

7 For two $n \times n$ matrices A and B one has $\det AB = \det A \cdot \det B$.

8 Matrix multiplication by hand is best organized by **Falk's scheme** as illustrated in Figure 1.1.

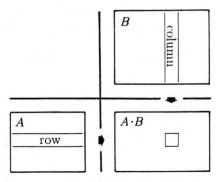

Figure 1.1: Falk's scheme.

9 Falk's scheme can also be used to procure an LU-factorization or a Cholesky factorization.

2 Linear Spaces

A **linear** or **vector space** $\mathbf{V}$ **over** $\mathbb{R}$ is a set which is closed under linear combinations with real coefficients. The elements of $\mathbf{V}$ are called **vectors**, the coefficients are **scalars**. A map from one linear space into another is called a **linear map** if it preserves linear combinations. The **standard vector space** is $\mathbb{R}^m$.

Literature: Greub, Strang, van der Waerden

2.1 Basis and Dimension

Let $\mathbf{o}$ denote the zero vector, then any r vectors $\mathbf{a}_1, \ldots, \mathbf{a}_r$ belonging to a vector space $\mathbf{V}$ are said to be **linearly dependent** if there exist scalars $x_1, \ldots, x_r$ not all of which are zero such that

$$\mathbf{a}_1 x_1 + \cdots + \mathbf{a}_r x_r = \mathbf{o} .$$

Otherwise $\mathbf{a}_1, \ldots, \mathbf{a}_r$ are said to be **linearly independent**. On building the matrix $A = [\mathbf{a}_1 \ldots \mathbf{a}_r]$, one has that $\mathbf{a}_1, \ldots, \mathbf{a}_r$ are linearly dependent if and only if $A\mathbf{x} = \mathbf{o}$ has a non-trivial solution.

The set of all linear combinations of the given $\mathbf{a}_i$ forms a linear space, called the **span** of the $\mathbf{a}_i$, or $span\,[\mathbf{a}_1 \ldots \mathbf{a}_r]$.

The space $\mathbf{A} = span\,[\mathbf{a}_1 \ldots \mathbf{a}_r]$ is called a **subspace** of $\mathbf{V}$. The **dimension** of $\mathbf{A}$, or $dim\,\mathbf{A}$, is defined as the maximum number of linearly independent vectors in $\mathbf{A}$. Occasionally, $n = dim\,\mathbf{A}$ is given as a superscript, $\mathbf{A}^n$

Let $\mathbf{a}_1, \ldots, \mathbf{a}_n$ be n linearly independent vectors of an n-dimensional linear space $\mathbf{V}$, and let $\mathbf{v}$ be some vector of $\mathbf{V}$. Then these $n + 1$ vectors are linearly dependent, i.e.,

$$\mathbf{v} = \mathbf{a}_1 x_1 + \cdots + \mathbf{a}_n x_n , \qquad x_i \in \mathbb{R} ,$$

in matrix notation $\mathbf{v} = A\mathbf{x}$. In this equation the factors x_i can be uniquely determined, otherwise the $\mathbf{a}_i$ would not be linearly independent. Hence $\mathbf{A}$ is non-singular. One says that the vectors $\mathbf{a}_1, \ldots, \mathbf{a}_n$ form a **basis** of $\mathbf{V}$. The $\mathbf{a}_i x_i$ are called the **components** of $\mathbf{v}$, while the x_i are referred to as the **coordinates** of $\mathbf{v}$ with respect to the $\mathbf{a}_i$.

Remark 1: It is convenient to denote a vector by the vector of its coordinates $\mathbf{x} = [x_1 \ldots x_n]^t$. This convention is used throughout this book.

Remark 2: On choosing some fixed basis of $\mathbf{V}^n$ every vector of $\mathbf{V}^n$ corresponds to a unique element of $\mathbb{R}^n$, and every linear combination in $\mathbf{V}^n$ corresponds to the same linear combination in $\mathbb{R}^n$. Therefore it is sufficient to consider $\mathbb{R}^n$ instead of $\mathbf{V}^n$. In particular, the $\mathbf{a}_i$ from above may be viewed as elements of $\mathbb{R}^m$, $m \geq n$.

2.2 Change of Bases

Let $\mathbf{a}_1, \ldots, \mathbf{a}_n$ and $\mathbf{b}_1, \ldots, \mathbf{b}_n$ denote two bases of a linear space $\mathbf{V}$. Then the $\mathbf{a}$'s can be expressed uniquely in terms of the $\mathbf{b}$'s,

$$\mathbf{a}_k = \mathbf{b}_1 c_{1,k} + \cdots + \mathbf{b}_n c_{n,k} .$$

Using matrix notation one gets

$$[\,\mathbf{a}_1 \quad \ldots \quad \mathbf{a}_n\,] = [\,\mathbf{b}_1 \quad \ldots \quad \mathbf{b}_n\,] \begin{bmatrix} c_{1,1} & \cdots & c_{1,n} \\ \vdots & & \vdots \\ c_{n,1} & \cdots & c_{n,n} \end{bmatrix} ,$$

or more concisely $A = BC$. As a consequence one has $C = B^{-1}A$, i.e., $C = [c_{i,k}]$ is non-singular since it is the product of non-singular matrices. Let $\mathbf{v}$ be some arbitrary vector of $\mathbf{V}$ with the representations $\mathbf{v} = A\mathbf{x} = B\mathbf{y}$, i.e.,

$$
\mathbf{v} = \begin{bmatrix} \mathbf{a}_1 & \cdots & \mathbf{a}_n \end{bmatrix} \begin{bmatrix} x_1 \\ \vdots \\ x_n \end{bmatrix} = \begin{bmatrix} \mathbf{b}_1 & \cdots & \mathbf{b}_n \end{bmatrix} \begin{bmatrix} y_1 \\ \vdots \\ y_n \end{bmatrix} .
$$

It then follows that $\mathbf{y} = C\mathbf{x}$, i.e.,

$$
\begin{bmatrix} y_1 \\ \vdots \\ y_n \end{bmatrix} = \begin{bmatrix} c_{1,1} & \cdots & c_{1,n} \\ \vdots & & \vdots \\ c_{n,1} & \cdots & c_{n,n} \end{bmatrix} \begin{bmatrix} x_1 \\ \vdots \\ x_n \end{bmatrix} .
$$

Note that the $\mathbf{a}$'s are expressed in terms of the $\mathbf{b}$'s, but the y's are expressed in terms of the x's. Both transformations are called **contragredient** to each other.

The representation $\mathbf{a}_k = B\mathbf{c}_k$ has a simple but important **geometric meaning**:

> The column $\mathbf{c}_k$ of C represents the coordinates of the basis vector $\mathbf{a}_k$ with respect to the basis $\mathbf{b}_1, \ldots, \mathbf{b}_n$.

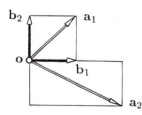

Example 1:

For $[\mathbf{a}_1\,\mathbf{a}_2] = [\mathbf{b}_1\,\mathbf{b}_2] \begin{bmatrix} 1 & 2 \\ 1 & -1 \end{bmatrix}$

one has $\quad \begin{bmatrix} y_1 \\ y_2 \end{bmatrix} = \begin{bmatrix} 1 & 2 \\ 1 & -1 \end{bmatrix} \begin{bmatrix} x_1 \\ x_2 \end{bmatrix} .$

2.3 Linear Maps

Of particular interest are maps which are compatible with the linear structure of linear spaces. Such maps must preserve linear combinations. Consider two linear spaces $\mathbf{A}$ and $\mathbf{B}$ with bases $\mathbf{a}_1, \ldots, \mathbf{a}_n$ and $\mathbf{b}_1, \ldots, \mathbf{b}_m$ respectively, and a map $\varphi : \mathbf{A} \to \mathbf{B}$ which preserves linear combinations, i.e.,

$$\varphi\left[\mathbf{a} \cdot \alpha + \mathbf{b} \cdot \beta\right] = \varphi\mathbf{a} \cdot \alpha + \varphi\mathbf{b} \cdot \beta$$

for all $\mathbf{a}, \mathbf{b} \in \mathbf{A}$ and all $\alpha, \beta \in \mathbb{R}$. Such a map φ is called a **linear map**.

The images of the $\mathbf{a}_k$ can uniquely be expressed in terms of the $\mathbf{b}$'s,

$$\varphi\mathbf{a}_k = \mathbf{b}_1 c_{1,k} + \cdots + \mathbf{b}_m c_{m,k} \ ,$$

which may be written in matrix notation as

$$\left[\,\varphi\mathbf{a}_1 \ \ldots \ \varphi\mathbf{a}_n\,\right] = \left[\mathbf{b}_1 \ldots \mathbf{b}_m\right] \begin{bmatrix} c_{1,1} & \cdots & c_{1,n} \\ \vdots & & \vdots \\ c_{m,1} & \cdots & c_{m,n} \end{bmatrix} ,$$

or in condensed form as $\varphi A = BC$. Let $\mathbf{a}$ be a vector of $\mathbf{A}$,

$$\mathbf{a} = \mathbf{a}_1 x_1 + \cdots + \mathbf{a}_n x_n = A\mathbf{x} \ ,$$

and $\mathbf{b} = \varphi\mathbf{a}$ its image in $\mathbf{B}$,

$$\mathbf{b} = \mathbf{b}_1 y_1 + \cdots + \mathbf{b}_m y_m = B\mathbf{y} \ .$$

Then one has $\mathbf{y} = B^{-1}\mathbf{b}$ and $\mathbf{b} = \varphi A\mathbf{x} = BC\mathbf{x}$. This implies $\mathbf{y} = C\mathbf{x}$, i.e.,

$$\begin{bmatrix} y_1 \\ \vdots \\ y_m \end{bmatrix} = \begin{bmatrix} c_{1,1} & \cdots & c_{1,n} \\ \vdots & & \vdots \\ c_{m,1} & \cdots & c_{m,n} \end{bmatrix} \begin{bmatrix} x_1 \\ \vdots \\ x_n \end{bmatrix} .$$

Note that the $\varphi\mathbf{a}$'s are expressed in terms of the $\mathbf{b}$'s via C, but the y's are expressed in terms of the x's, i.e., both transformations are **contragredient** to each other.

The representation $\varphi\mathbf{a}_k = B\mathbf{c}_k$ has a simple but important **geometric meaning**:

> The column $\mathbf{c}_k$ of C represents the coordinates of the image $\varphi\mathbf{a}_k$ of the basis vector $\mathbf{a}_k$ with respect to the basis $\mathbf{b}_1, \ldots, \mathbf{b}_m$.

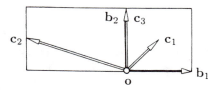

Example 2: On inspecting the figure one obtains the matrix C,

$$[\mathbf{c}_1\ \mathbf{c}_2\ \mathbf{c}_3] = \begin{bmatrix} 1/2 & -3/2 & 0 \\ 1/2 & 1/2 & 1 \end{bmatrix}.$$

2.4 Kernel and Fibers

The images $\varphi\mathbf{a}_k$ span $\varphi\mathbf{A}$, the **image** of $\mathbf{A}$. The image of $\mathbf{A}$ is a subspace of $\mathbf{B}$ with $dim\ \varphi\mathbf{A} \le dim\ \mathbf{A}$. These dimensions can be analyzed in more detail. There exists a subspace $\mathbf{K} \subset \mathbf{A}$, called the **kernel** of φ, $\mathbf{K} = kern\ \varphi$, which is the set of all vectors in $\mathbf{A}$ mapped into the null vector $\mathbf{o}$ of $\mathbf{B}$. The subspace $\mathbf{K}$ is represented by the solution of the homogeneous system

$$C\mathbf{x} = \mathbf{o}\ .$$

For any fixed vector $\mathbf{a}$ of $\mathbf{A}$ and all elements $\mathbf{k}$ of this kernel $\mathbf{K}$, the subset $\mathcal{F}_a$ of $\mathbf{A}$ formed by all $\mathbf{a} + \mathbf{k}$ is called the **fiber** over $\mathbf{a}$. Evidently, φ maps all elements of $\mathcal{F}_a$ into the same image $C\mathbf{a}$. Note that a fiber is a linear space only if $\mathbf{a} = \mathbf{o}$.

Using a basis of $\mathbf{A}$ which contains a basis of $\mathbf{K}$ one finds that

$$dim\ \varphi\mathbf{A} + dim\ \mathbf{K} = dim\ \mathbf{A}\ .$$

Example 3: In Example 2, $\mathbf{K}$ consists of all vectors $[3\ \ 1\ -2]^t\lambda$, with $\lambda \in \mathbb{R}$.

2.5 Point Spaces

One can see the world as a **space of points**. This point space is closely
related to a linear space in a natural way. Two points are connected by a
vector and a vector added to a point gives a point again. These relations
are expressed by the notation

$$\mathbf{v} = \mathbf{p} - \mathbf{a} \qquad \text{and} \qquad \mathbf{p} = \mathbf{a} + \mathbf{v} \,,$$

where $\mathbf{a}$ and $\mathbf{p}$ are points and $\mathbf{v}$ is the vector pointing from $\mathbf{a}$ to $\mathbf{p}$. Let $\mathbf{v}$
be given with respect to a basis of $\mathbf{A}^n$, $\mathbf{v} = \mathbf{a}_1 x_1 + \cdots + \mathbf{a}_n x_n$, then

$$\mathbf{p} = \mathbf{a} + \mathbf{a}_1 x_1 + \cdots + \mathbf{a}_n x_n \,.$$

Let $\mathbf{a}$ be a fixed point, then every coordinate column $\mathbf{x} = [x_1 \ldots x_n]^t$
defines a point $\mathbf{p}$, with different $\mathbf{x}$'s generating different points.

Affine spaces: The set of points $\mathbf{p}$ corresponding to all $\mathbf{x} \in \mathbb{R}^n$ is called
an **affine space** $\mathcal{A}$, while $span\,[\mathbf{a}_1 \ldots \mathbf{a}_n]$ is called the **underlying vector
space A**. One defines $dim\,\mathcal{A} = dim\,\mathbf{A}$. A point $\mathbf{a} \in \mathcal{A}$ together with a
basis $\mathbf{a}_1, \ldots, \mathbf{a}_n \in \mathbf{A}$ form an **affine system** in $\mathcal{A}$. The point $\mathbf{a}$ is referred
to as the **origin** while the x_i are called the **affine coordinates** of $\mathbf{p}$ with
respect to the frame $\mathbf{a}; \mathbf{a}_1, \ldots, \mathbf{a}_n$. Affine spaces are discussed in Part III.

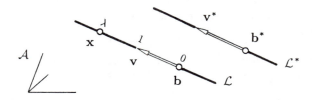

Figure 2.1: Parallelism and affine scale.

In most parts of this book, when points are viewed as vectors they will be
denoted by their coordinate columns $\mathbf{x}$ with respect to some fixed frame.
Note that $\mathbf{p}$ and $\mathbf{x}$ above denote the same point with respect to different
systems.

An **affine subspace** $\mathcal{S}$ of $\mathcal{A}$ is defined by some point $\mathbf{b} \in \mathcal{A}$ and a subspace $\mathbf{S}$ of $\mathbf{A}$, i.e., $\mathcal{S} = \{\mathbf{b} + \mathbf{v} \mid \mathbf{v} \in \mathbf{S}\}$. In particular, a line $\mathcal{L}$ is a 1-dimensional subspace. It will be represented as

$$\mathbf{x} = \mathbf{b} + \mathbf{v}\lambda \ .$$

The parameter λ is called an **affine scale** on $\mathcal{L}$. It represents $\mathbf{x}$ with respect to the affine system $\mathbf{b}; \mathbf{v}$. Using this scale, the **ratio** of the point λ with respect to the points λ_0 and λ_1 is defined by

$$ratio\,(\lambda; \lambda_0, \lambda_1) = \frac{\lambda - \lambda_0}{\lambda_1 - \lambda} \ .$$

Note that this ratio depends on the ordering of the points, but not on the respective affine scale.

The line $\mathcal{L}$ is said to be **parallel** to a second line $\mathcal{L}^*$ given by

$$\mathbf{x} = \mathbf{b}^* + \mathbf{v}^*\mu$$

if $\mathbf{v} = \mathbf{v}^*\sigma$, $\sigma \neq 0$.

Euclidean spaces: If the basis vectors $\mathbf{a}_i$ of the underlying vector space $\mathbf{A}$ have length 1 and are pairwise perpendicular, then the corresponding affine system is called a **Cartesian system**. The x_i are called **Cartesian coordinates**, while the space $\mathcal{A}$ is called a **Euclidean space** and denoted by $\mathcal{E}$. In a Cartesian system the square of the distance between two points $\mathbf{x}$ and $\mathbf{x} + \mathbf{d}$ equals $\mathbf{d}^t\mathbf{d}$, and two vectors $\mathbf{u}$ and $\mathbf{v}$ are perpendicular if $\mathbf{u}^t\mathbf{v} = 0$. Euclidean spaces are discussed in Part IV.

Projective spaces: Often it is easier to describe geometric properties if one introduces points at infinity — one point for each 1-dimensional subspace of $\mathbf{A}$. Then any two parallel lines meet in a point at infinity. These points are called **ideal points**, while the 1-dimensional subspaces of $\mathbf{A}$ are called **directions** of $\mathcal{A}$. The ideal points of $\mathcal{A}$ form the **ideal hyperplane** $\mathcal{A}_\infty$ of $\mathcal{A}$. The union $\mathcal{P} = \mathcal{A} \cup \mathcal{A}_\infty$ is called the **projective extension** of $\mathcal{A}$. It represents the prototype of a **projective space**. Projective spaces are discussed in Part V.

2.6 Notes and Problems

1 Although the elements of $\mathbb{R}^n$ can be interpreted as the elements of either an affine space or a linear space, the structures of these spaces are different.

2 The solution of a homogeneous linear system forms a linear space.

3 The solution of a non-homogeneous linear system forms an affine space.

4 More exactly, any r independent linear equations in n variables define an affine space of dimension $n - r$, provided that the corresponding linear system has a solution.

5 Any linear space is in a natural way an affine space, but not vice versa.

6 The set theoretical intersection of two subspaces $\mathbf{A}$ and $\mathbf{B}$ of a linear space is a linear space and is called the **intersection** $\mathbf{A} \sqcap \mathbf{B}$ of $\mathbf{A}$ and $\mathbf{B}$.

7 The set theoretical **union** of two subspaces $\mathbf{A}$ and $\mathbf{B}$ of a linear space is a linear space only if $\mathbf{A} \subset \mathbf{B}$ or $\mathbf{B} \subset \mathbf{A}$.

8 Let $\mathbf{A} = span\,[\mathbf{a}_1 \ldots \mathbf{a}_r]$ and $\mathbf{B} = span\,[\mathbf{a}_{r+1} \ldots \mathbf{a}_s]$, then $span\,[\mathbf{a}_1 \ldots \mathbf{a}_s]$ is called the **join** $\mathbf{A} \sqcup \mathbf{B}$ of $\mathbf{A}$ and $\mathbf{B}$.

9 Let $\mathbf{a}_1, \ldots, \mathbf{a}_r$ be linearly independent vectors of some n-dimensional linear space $\mathbf{A}$. They can be supplemented to a basis $\mathbf{a}_1, \ldots, \mathbf{a}_n$ of $\mathbf{A}$.

10 Given r non-zero but linearly dependent vectors $\mathbf{a}_1, \ldots, \mathbf{a}_r$, one can construct a basis of $span\,[\mathbf{a}_1 \ldots \mathbf{a}_r]$ by the Gauss-Jordan algorithm.

11 The set of all one-dimensional subspaces of a linear space $\mathbf{V}$ forms a **projective space** $\mathcal{P}$.

3 Least Squares

A linear system is **overdetermined** if the number of equations exceeds the number of unknowns. Since such a system has no solution in general, one usually seeks unknowns which "solve "the system best, approximatively. Frequently, one minimizes some Euclidean distance. This concept leads to the **method of least squares**.

Literature: Boehm·Prautzsch, Conte·de Boor, Wilkinson

3.1 Overdetermined Systems

Let A be a tall $m \times n$ matrix, i.e., $m > n$, with $rank\, A = n$, and let $Ax = \mathbf{a}$ be a linear system

Only if $\mathbf{a}$ is a linear combination of the columns $\mathbf{a}_k$ of A, is there a solution $\mathbf{x}$. But, in general, one has

$$\mathbf{r} = A\mathbf{x} - \mathbf{a} \neq \mathbf{o} \qquad \text{for all } \mathbf{x} \in \mathbb{R}^n \ .$$

The column $\mathbf{r}$ is called the **residual vector** associated with $\mathbf{x}$. It can be interpreted in $\mathbb{R}^m$ as the vector from the point $\mathbf{a}$ to the point $A\mathbf{x}$ as illustrated in Figure 3.1. An approximate solution $\mathbf{x}$ which minimizes $\mathbf{r} = \mathbf{r}(\mathbf{x})$ in some sense is all one can hope for. Minimizing $\mathbf{r}^t\mathbf{r}$ is rather a simple task.

In the Euclidean space $\mathcal{E}^m$ the length of $\mathbf{r}$ is minimal if $\mathbf{r}$ is orthogonal to the subspace $\mathcal{A}$ spanned by $\mathbf{o}; \mathbf{a}_1, \ldots, \mathbf{a}_n$, i.e., if

$$A^t\mathbf{r} = \mathbf{o} .$$

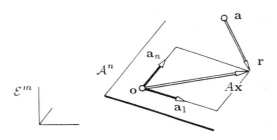

Figure 3.1: The residual vector.

Substituting $A\mathbf{x} - \mathbf{a}$ for $\mathbf{r}$ results in the so-called **Gaussian normal equations**,

$$A^t A\mathbf{x} = A^t\mathbf{a} .$$

The solution $\mathbf{x}$ represents the **foot** of the perpendicular from $\mathbf{a}$ onto $\mathcal{A}$ with respect to the affine system $\mathbf{o}; \mathbf{a}_1, \ldots, \mathbf{a}_n$. Note that $A^t A$ is an $n \times n$ matrix and $A^t\mathbf{a}$ is an n column. Moreover, $A^t A$ is symmetric and, if the $\mathbf{a}_i$ are linearly independent, also positive definite. In this case the normal equations can be solved via a symmetric factorization of $A^t A$, as mentioned in Remark 5 of Section 1.5.

Remark 1: In general, normal equations are poorly conditioned. Hence, it is advisable to use a numerically stable method such as Householder's. In **Householder's method**, the matrix $[A|\mathbf{a}]$ is multiplied by a sequence H of orthonormal transformations to obtain a matrix $[B|\mathbf{b}]$ such that B is composed of an upper triangular matrix U and a null matrix O,

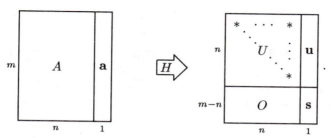

Since orthonormal transformations do not change the Euclidean length of a vector, $\mathbf{r} = A\mathbf{x} - \mathbf{a}$ and $H\mathbf{r} = B\mathbf{x} - \mathbf{b}$ have the same length, i.e., the solution of $U\mathbf{x} = \mathbf{u}$ minimizes $\mathbf{r}^t\mathbf{r}$, where $\mathbf{s}^t\mathbf{s}$ is the minimum value of $\mathbf{r}^t\mathbf{r}$.

Remark 2: The individual equations of $A\mathbf{x} = \mathbf{a}$ may be multiplied with arbitrary weights. This "scaling" changes the coordinates of the residual vector and, hence, influences the result. Thus one may distinguish equations corresponding to very accurate measurements. In this way, equations stemming from accurate measurements can become more influential than others.

3.2 Homogeneous Systems

The least squares method fails for homogeneous systems, i.e., if $\mathbf{a} = \mathbf{o}$, because $\mathbf{x} = \mathbf{o}$ solves the system. A simple way to avoid this problem is to add a constraint by setting one of the x_k's equal to 1. On constraining, e.g., x_1, one has to "solve" the overdetermined non-homogeneous system

$$m \left[\; \mathbf{a}_2 \;\cdots\; \mathbf{a}_n \;\right] \begin{bmatrix} x_2 \\ \vdots \\ x_n \end{bmatrix} = - \begin{bmatrix} \mathbf{a}_1 \end{bmatrix} .$$

$n-1$

Obviously, the "solution" depends on which coordinate x_k is constrained. Note that the corresponding $\mathbf{a}_k$ must be distinctly different from $\mathbf{o}$ to avoid numerical instabilities.

3.3 Constrained Least Squares

Sometimes the "solution" of an overdetermined system $A\mathbf{x} = \mathbf{a}$ is required
to satisfy an additional system $B\mathbf{x} = \mathbf{b}$, $\mathbf{b} \neq \mathbf{o}$. One can compute the
solution of the system $B\mathbf{x} = \mathbf{b}$ by the Gauss-Jordan algorithm and obtain
an equivalent system

$$\mathbf{x} = \mathbf{c} + C\mathbf{t} .$$

These additional constraints are **hard constraints** which could be, for example, the boundary conditions of some initial problem. On substituting
$\mathbf{c} + C\mathbf{y}$ for $\mathbf{x}$, the initial system $A\mathbf{x} = \mathbf{a}$ reduces to the overdetermined
system

$$AC\mathbf{y} = \mathbf{a} - A\mathbf{c} .$$

If A is an $m \times n$ matrix and B is an $l \times n$ matrix, then AC is an $m \times n\text{-}l$
matrix. Note that $m > n > l$.

Geometrically, this procedure may be interpreted as the restriction of $\mathbf{x}$
to a subspace $\mathcal{C}$ of $\mathcal{A}$ and the introduction of new affine coordinates in $\mathcal{C}$
represented by $\mathbf{y}$. Note that this method works even if $\mathbf{a} = \mathbf{o}$.

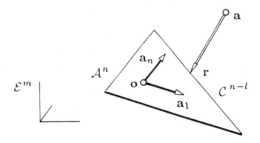

Figure 3.2: Constrained least squares.

Example 1: An example is discussed in Section 3.2 where $x_1 = 1$ represents the hard constraint, i.e.,

$$\mathbf{c} = \begin{bmatrix} 1 \\ 0 \\ \vdots \\ 0 \end{bmatrix} \quad \text{and} \quad C = \begin{bmatrix} 0 & \cdots & 0 \\ 1 & & \\ & \ddots & \\ & & 1 \end{bmatrix} .$$

3.4 Linearization

One can use the above methods to iteratively approximate zeros of a system of m non-linear equations

$$f_1(u_1, \ldots, u_n) = 0$$

$$\vdots$$

$$f_m(u_1, \ldots, u_n) = 0 \ ,$$

in n variables u_k, in vector form $\mathbf{f}(\mathbf{u}) = \mathbf{o}$. Let

$$J_f(\mathbf{u}) = \left[\frac{\partial f_i}{\partial u_k} \right]_{\mathbf{u}}$$

denote the $m \times n$ **Jacobi-matrix** of $\mathbf{f}$ at $\mathbf{u}$. Then the Taylor expansion of $\mathbf{f}$ at $\mathbf{u}$ is given by

$$\mathbf{f}(\mathbf{u} + \triangle\mathbf{u}) = \mathbf{f}(\mathbf{u}) + J_f(\mathbf{u}) \cdot \triangle\mathbf{u} + \cdots \ .$$

The first two terms of this expansion provide an approximation of $\mathbf{f}(\mathbf{u}+\triangle\mathbf{u})$ which is linear in $\triangle\mathbf{u}$. It is the so-called **linearization** at $\mathbf{u}$ of the equation $\mathbf{f} = \mathbf{o}$, namely

$$\mathbf{f}(\mathbf{u}) + J_f(\mathbf{u}) \cdot \triangle\mathbf{u} = \mathbf{o} \ .$$

Let $\mathbf{u}$ be a first approximation of a root of $\mathbf{f}$ and let $\triangle\mathbf{u}$ solve the linearization at $\mathbf{u}$; then $\mathbf{u} + \triangle\mathbf{u}$ is often an improved approximation.

Remark 3: If $\mathbf{f}(\mathbf{u}) = \mathbf{o}$ is an overdetermined system, one can solve the linearization of $\mathbf{f} = \mathbf{o}$ by the method of Section 3.1.

Example 2: Given a curve $\mathbf{x}(u)$ and a point $\mathbf{p}$ in $\mathcal{E}^m$, one wants to find a point $\mathbf{x}$ on the curve with minimum distance to $\mathbf{p}$. From $\mathbf{f}(u) = \mathbf{x}(u) - \mathbf{p} = \mathbf{o}$ one obtains both the linearization at a first approximation of the wanted u

$$\mathbf{x} - \mathbf{p} + \dot{\mathbf{x}}\triangle u = \mathbf{o} \ ,$$

consisting of m linear equations in the only variable u, and the Gaussian normal equation

$$\dot{\mathbf{x}}^t \left[\mathbf{x} - \mathbf{p}\right] + \dot{\mathbf{x}}^t \dot{\mathbf{x}}\triangle u = 0 \ ,$$

where the dot denotes differentiation with respect to u.

Example 3: Given a surface $\mathbf{x}(u,v)$ and a point $\mathbf{p}$ in $\mathcal{E}^m$, one wants a point $\mathbf{x}$ on the surface with minimum distance to $\mathbf{p}$. From the equation $\mathbf{f}(u,v) = \mathbf{x}(u,v) - \mathbf{p} = \mathbf{o}$ one obtains the linearization at a first approximation of the wanted (u,v)

$$\mathbf{x} - \mathbf{p} + [\mathbf{x}_u \mathbf{x}_v]\, \triangle \mathbf{u} = \mathbf{o} ,$$

consisting of m equations in the two variables u and v, and the Gaussian normal equations

$$[\mathbf{x}_u \mathbf{x}_v]^t\,[\mathbf{x} - \mathbf{p}] + [\mathbf{x}_u \mathbf{x}_v]^t\,[\mathbf{x}_u \mathbf{x}_v]\, \triangle \mathbf{u} = \mathbf{o} ,$$

where $\mathbf{x}_u$ and $\mathbf{x}_v$ denote partial derivatives.

3.5 Underdetermined Systems

If A is a wide $m \times n$ matrix, i.e., if $m < n$, then the linear system $A\mathbf{x} = \mathbf{a}$ is **underdetermined**, and in general, it has more than one solution. One is interested in a solution which is close in some sense to a given n column $\mathbf{p}$. Following the ideas from Section 3.1 and as illustrated in Figure 3.3 in $\mathcal{E}^n$, the Euclidean distance between $\mathbf{x}$ and $\mathbf{p}$ will be minimal if $\mathbf{x} - \mathbf{p}$ is a linear combination of the columns of A^t, i.e., $\mathbf{x} - \mathbf{p} = A^t\mathbf{y}$, where $\mathbf{y}$ is an unknown m column. Substituting $\mathbf{x}(\mathbf{y})$ into $A\mathbf{x} = \mathbf{a}$ one gets a linear system for $\mathbf{y}$

$$AA^t\mathbf{y} = \mathbf{a} - A\mathbf{p} ,$$

where AA^t is a symmetric $m \times m$ matrix and $\mathbf{a} - A\mathbf{p}$ is an m column. These equations are also called **Gaussian normal equations**, while

$$\mathbf{x} = \mathbf{p} + A^t\mathbf{y}$$

is called the **correlator equation**.

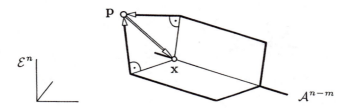

Figure 3.3: Underdetermined system.

3.6 Notes and Problems

1 One can use a **linear program**, as considered in Operations Research, to minimize the maximum component of the residual vector of an overdetermined system.

2 Finding the foot point of the perpendicular from **a** to a subspace $\mathcal{A}$ which is given by a set of linear equations as discussed in Section 3.5 is called the **dual fitting problem**. It is dual to the method discussed in Section 3.1.

3 The improvement of an approximation by alternating the linearization of **f** and the solving of the corresponding system is known as the **Newton-Raphson method**.

4 Geometrically, linearization corresponds to the replacement of each surface $f_i(\mathbf{x}) = constant$ by its $(n-1)$-dimensional tangent plane at **u**.

5 The Gaussian normal equations from Example 2 and Remark 3 above allow a simple geometric interpretation as illustrated in Figure 3.4: The point $\mathbf{x} + \triangle\mathbf{x}$ represents the foot from **p** to the tangent or tangent plane at the point **x**.

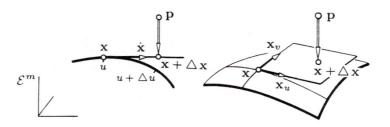

Figure 3.4: Linearization, examples.

6 Linearization is also used in marching methods to compute the intersection of two surfaces as illustrated in Figure 3.5. In each step of a
marching method one replaces both surfaces by their tangent planes at
some points **x** and **y** and then chooses a point **p** on the intersection
line. One then uses the foot points of the perpendiculars from **p** to the
surfaces to update **x** and **y** for the next step of the marching.

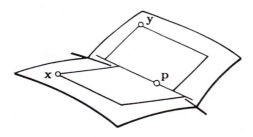

Figure 3.5: Principle of the marching method.

PART TWO

Images and Projections

Technical and scientific illustrations are supposed to convey exact information as to the shape and size of certain objects. For this reason physical objects are most commonly represented by their projections into a plane using either parallel rays or a family of rays emanating from a center. While central projections can be realized physically by a camera, parallel projections can not. Regardless, both kinds of images·can be constructed directly without a physical device. Moreover, one can reconstruct the shape and size of an object from a pair of such projections.

4 Parallel Projections

Although parallel projections cannot be realized physically, the concept of parallel rays has several advantages. Parallel projections are simple to construct and allow for an easy reconstruction of the measurements of an object from its image. Convincing examples of parallel projections include the military and cavalier projections which were first used by G. Monge (1746–1818) to construct scaled drawings of fortifications.

Literature: Hohenberg, Penna·Patterson, Rehbock

4.1 Pohlke's Theorem

It is intriguing to observe that a parallel projection of two parallel lines produces a pair of parallel lines and that two parallel distances have the same ratio as their two parallel images. These properties, which are illustrated in Figure 4.1, are summarized in the **fundamental theorem**:

Parallel projections preserve parallelism and ratios.

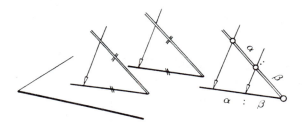

Figure 4.1: Parallelism and ratio.

On employing just these two simple properties as **drawing rules**, one can draw any parallel projection. This is a consequence of a theorem given by Pohlke in 1853 and illustrated below. Pohlke's Theorem asserts that a two-dimensional figure of a cube is a parallel projection of a cube if parallel edges have parallel images. In other words, a cube of the right size can be positioned in space and projected by parallel rays such that its image coincides with the drawn figure.

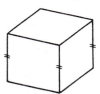

Pohlke's Theorem: Any figure like the one on the left exhibits the parallel projection of a cube.

More generally any **parallel projection** can be obtained by means of these two rules. The principle of the procedure is illustrated in Figure 4.2 with a very simple object. The object itself is given by two scaled elevations.

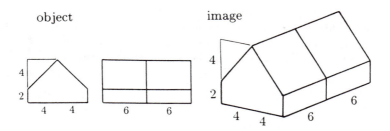

object image

Figure 4.2: The principle of drawing a parallel projection.

Example 1: Most illustrations in this book are parallel projections.

Example 2: One can extend the top view of an object to a parallel projection such that vertical distances are preserved. The resulting projection leaves horizontal cross sections invariant. Figure 4.3 shows an example of this so-called **military projection**.

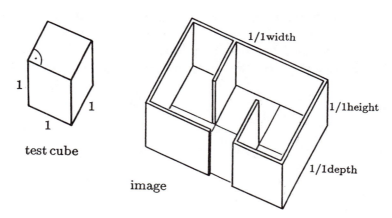

test cube

image

Figure 4.3: Military projection.

Example 3: One can extend the front view of an object to a parallel projection such that lines orthogonal to the front plane are projected onto lines with a slope of 45° while the distances on these lines are halved. The resulting projection preserves all cross sections parallel to the front plane. Figure 4.4 shows an example of this so-called **cavalier projection**.

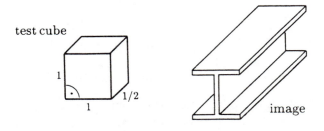

Figure 4.4: Cavalier projection.

Remark 1: The **quality** of a projection, i.e., its visual appeal, can be tested by drawing the corresponding unit cube.

Remark 2: A proof of Pohlke's theorem, i.e., the construction of position and size of the cube in space, is not simple. However, this construction is quite evident for military and cavalier projections, as illustrated in Figure 4.5, where the shadows show these special projections.

Military projection   Cavalier projection

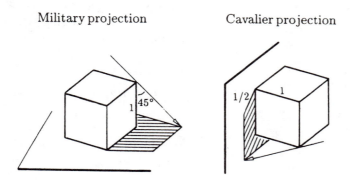

Figure 4.5: Simple examples of Pohlke's theorem.

4.2 Orthogonal Projections

A projection whose rays are perpendicular to the image plane is called an **orthogonal projection**. In general, projections are not orthogonal. Consequently the image of a sphere is generally an ellipse; it is a circle only in the special case when the projection is orthogonal.

The orthogonal projections can be distinguished among the parallel projections by certain characteristic conditions. Imagine some test cube with edge length a. Consider the images of the three edges incident at a corner of the cube and denote their lengths by a_1, a_2, a_3 respectively. Let $\gamma_1, \gamma_2, \gamma_3$ be the angles as shown in the figure below. By some trigonometrical considerations one can derive the following conditions:

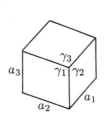

$$a_1^2 = -\frac{\cos\gamma_1}{\sin\gamma_2 \sin\gamma_3}a^2, \dots$$

$$\sin^2\gamma_1 = \frac{a^2 - a_1^2}{a_2^2 a_3^2}a^2, \dots \ .$$

Note that $\gamma_1 + \gamma_2 + \gamma_3 = 360°$ and $a_1^2 + a_2^2 + a_3^2 = 2a^2$.

Example 4: Special orthogonal projections are the **plan view**, **elevation**, and **side elevation** defined by

$$a_3 = 0, a_1 = a_2 = a,$$
$$a_1 = 0, a_2 = a_3 = a \text{ , and}$$
$$a_2 = 0, a_3 = a_1 = a,$$

respectively. See also Figure 4.2 for another example.

Example 5: Requiring $a_1 : a_2 : a_3 = 1 : 1 : 1$ (as in the military projection) gives

$$\gamma_1 = \gamma_2 = \gamma_3 = 120° \text{ ,}$$
$$a_1^2 = a_2^2 = a_3^2 = \tfrac{2}{3}a^2 \text{ .}$$

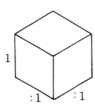

Obviously, this so-called **isometric projection** is orthogonal because of its symmetry.

Example 6: Requiring $a_1 : a_2 : a_3 = \tfrac{1}{2} : 1 : 1$ (as in the cavalier projection) gives

$$a_1^2 = \tfrac{2}{9} a^2 \text{ ,} \qquad a_2^2 = a_3^2 = \tfrac{8}{9} a^2 \text{ ,}$$
$$\gamma_1 = 97.18° \text{ ,} \qquad \gamma_2 = \gamma_3 = 131.41° \text{ .}$$

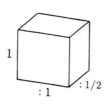

This orthogonal and so-called **dimetric projection** has many nice properties, e.g., $\cos \gamma_1 = -\tfrac{1}{8}$, $\sin \gamma_2 = \tfrac{6}{8}$. It has been used especially for hand drawings and found its way into several industry standards.

Remark 3: Methods to generate arbitrary orthogonal projections are discussed in Chapters 5 and 24.

4.3 Computing a Parallel Projection

Two Cartesian systems are involved in computing a projection: Let $\mathbf{x} = [x\ y\ z]^t$ be some object point given with respect to the object system and let $\mathbf{y} = [\xi\ \eta]^t$ be its image with respect to the image system.

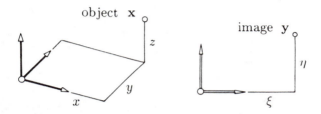

Figure 4.6: Object and image system.

From the fundamental theorem in Section 4.1 one can conclude that

$$\mathbf{y} = \mathbf{a} + \mathbf{a}_1 x + \mathbf{a}_2 y + \mathbf{a}_3 z \ ,$$

where $\mathbf{a}$ represents the image of the origin and $\mathbf{a}_k$ the image of the kth basis vector of the object system. More concisely one has

$$\mathbf{y} = \mathbf{a} + A\mathbf{x}, \qquad \text{where} \quad A = [\mathbf{a}_1\ \mathbf{a}_2\ \mathbf{a}_3]\ .$$

Note that the object system is represented with respect to itself by $\mathbf{o}$ and the three columns of the identity matrix.

Furthermore, the difference $\triangle\mathbf{x}$ of two points $\mathbf{x}$ and $\mathbf{x} + \triangle\mathbf{x}$ has the image

$$\triangle\mathbf{y} = A\triangle\mathbf{x}\ .$$

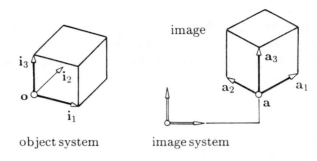

Figure 4.7: The image of the object system.

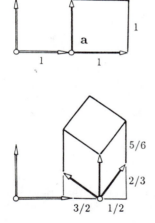

Example 7: The elevation (front view) as defined in Section 4.2 is described by

$$\mathbf{a} = \begin{bmatrix} 1 \\ 0 \end{bmatrix}, \qquad A = \begin{bmatrix} 0 & 1 & 0 \\ 0 & 0 & 1 \end{bmatrix}.$$

Example 8: An example of the military projection is given by

$$\mathbf{a} = \begin{bmatrix} 3/2 \\ 0 \end{bmatrix}, \qquad A = \begin{bmatrix} 1/2 & -2/3 & 0 \\ 2/3 & 1/2 & 5/6 \end{bmatrix}.$$

Note that A includes a scaling by the factor $5/6$.

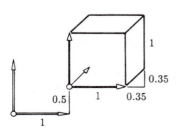

Example 9: An example of the cavalier projection is given by

$$\mathbf{a} = \begin{bmatrix} 1 \\ 0.5 \end{bmatrix}, \qquad A = \begin{bmatrix} 0.35 & 1 & 0 \\ 0.35 & 0 & 1 \end{bmatrix}.$$

4.4 Projecting Rays

Although the position and size of the object in space are not easy to re-construct from the object's image, the direction of the projecting is. If two points have the same image, their difference $\triangle\mathbf{x}$ is mapped onto $\mathbf{o}$, i.e., one has

$$A\triangle\mathbf{x} = \mathbf{o} .$$

The solutions $\triangle\mathbf{x}$ of this homogeneous system, representing the direction of the projecting rays, can be written down using Cramer's rule.

Example 10: For the projection discussed in Example 8 one gets that $\triangle\mathbf{x}^t = \lambda\,[0.4 \quad 0.3 \quad -0.5]$, i.e., the direction of the projecting rays is given by $[4 \quad 3 \quad -5]^t$.

Example 11: For the projection discussed in Example 9 one gets that $\triangle\mathbf{x}^t = \mu[1 \quad -.35 \quad -.35]$, i.e., the direction of the projection rays is given by $[20 \quad -7 \quad -7]^t$.

4.5 Notes and Problems

1 A circular disk parallel to the image plane will always be projected onto a circular disk, whether the projection rays are orthogonal to the image plane or not.

2 A circle orthogonal to the projection rays will be projected onto an ellipse. The ratio a/b of the axes equals $\sin\delta$, where δ is the angle between the rays and the image plane. Moreover, a equals the radius of the circle.

3 As a consequence of Note 2 the military projection of a sphere of radius r is an ellipse with semi-axes $\sqrt{2}r$ and r. Note that the principal axis of the ellipse is parallel to the image of the z-axis.

4 The military projection of a side of a cube parallel to the coordinate axes forms a rhombus. Consequently, the images of the diagonals are orthogonal.

5 As a consequence of Note 4, any circle on one of the sides of such a cube is projected into an ellipse such that each axis is parallel to a diagonal.

6 The orthogonal projection of a circular disk of radius r is an ellipse whose semi-axes have lengths r and $r \cdot \cos \varepsilon$, where ε is the angle between the plane of the disk and the image plane.

7 The parallel projection of a sphere is a circle only if the projection is orthogonal. This may be used to check whether a given projection is orthogonal.

8 The orthogonal projection of a right angle is a right angle only if one of its legs is parallel to the image plane.

9 Every map $\mathbf{y} = \mathbf{a} + A\mathbf{x}$ from 3-space onto a plane represents a parallel projection composed with a scaling.

10 Any parallel projection of a parallel projection is a scaled parallel projection.

11 In order to make the relation between top, front and side view clear these elevations are usually combined in one drawing with common axes as illustrated in Figure 4.8.

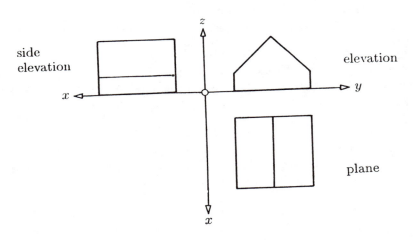

Figure 4.8: Composing elevations.

5 Moving the Object

The previous methods are concerned with the projection of a fixed object into a fixed image plane. One cannot use them to compute or to illustrate either a controlled motion of an object or a controlled view by a continuously moving observer. In order to obtain these images one has to compose motions with projections. A matrix representation facilitates the computation of such compositions.

Literature: Hearn, Mortenson, Penna·Patterson

5.1 Euclidean Motions

Every motion in a Euclidean space can be understood as a special affine map. Let $\mathbf{x}$ and $\mathbf{y}$ denote the respective positions of some object point before and after a motion in 3-space. Then $\mathbf{y}$ can be written as

$$\mathbf{y} = \mathbf{b} + \mathbf{b}_1 x + \mathbf{b}_2 y + \mathbf{b}_3 z \ ,$$

or concisely as

$$\mathbf{y} = \mathbf{b} + B\mathbf{x} \ ,$$

where $\mathbf{b}$ and the columns $\mathbf{b}_1, \mathbf{b}_2, \mathbf{b}_3$ of B represent the moved coordinate system with respect to the system in its initial position. Note that the $\mathbf{b}$'s and $\mathbf{y}$ are 3-columns. Since the map represents a motion in Euclidean space, it preserves the Euclidean distance between any two points. Let $\triangle \mathbf{x}$

and $\triangle\mathbf{y}$ denote the differences between the points $\mathbf{x}_1, \mathbf{x}_2$ and $\mathbf{y}_1, \mathbf{y}_2$. Then, one has

$$\triangle\mathbf{y}^t\triangle\mathbf{y} = \triangle\mathbf{x}^t B^t B\triangle\mathbf{x} = \triangle\mathbf{x}^t\triangle\mathbf{x}, \qquad \text{for all } \triangle\mathbf{x} .$$

Hence $B^t B = I$, i.e., B is **orthonormal**. Additionally, one has $\det B = 1$ since a motion does not change the orientation of the coordinate system. Figure 5.1 illustrates this motion, where the object simply is the object system. Note that the results hold true only if the coordinate system is Cartesian.

Every orthogonal projection can be realized by a motion followed by a side elevation which sets $\eta = 0$, where η denotes the 2nd coordinate of $\mathbf{y}$.

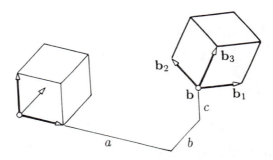

Figure 5.1: Motion of the system.

Example 1: An initial scaling by a positive factor ϱ, followed by a translation by a in the 1-direction is given through

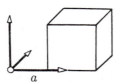

$$\mathbf{b} = \begin{bmatrix} a \\ 0 \\ 0 \end{bmatrix}, \quad B = \begin{bmatrix} \varrho & 0 & 0 \\ 0 & \varrho & 0 \\ 0 & 0 & \varrho \end{bmatrix} .$$

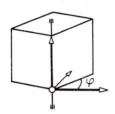

Example 2: A rotation around the 3-axis by φ without translation is given through

$$\mathbf{b} = \begin{bmatrix} 0 \\ 0 \\ 0 \end{bmatrix}, \quad B_\varphi = \begin{bmatrix} \cos\varphi & -\sin\varphi & 0 \\ \sin\varphi & \cos\varphi & 0 \\ 0 & 0 & 1 \end{bmatrix}.$$

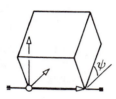

Example 3: A rotation around the 1-axis by ψ without translation is given through

$$\mathbf{b} = \begin{bmatrix} 0 \\ 0 \\ 0 \end{bmatrix}, \quad B_\psi = \begin{bmatrix} 1 & 0 & 0 \\ 0 & \cos\psi & -\sin\psi \\ 0 & \sin\psi & \cos\psi \end{bmatrix}.$$

5.2 Composite Motions

Any motion can be composed of the three motions in Examples 1, 2, and 3 above. The following example serves as a demonstration. The motion illustrated in Figure 5.2 is composed of:

1st   a rotation around the 3-axis by φ,
2nd   a rotation around the 1-axis by ψ, and
3rd   a translation by $\mathbf{b}$.

Finally one may set $\eta = 0$ in order to obtain an orthogonal projection. To simplify the figure, the 2nd coordinate is generally suppressed, i.e., only side elevations are shown in the figure. In matrix notation the total motion is of the form

$$\mathbf{y} = \mathbf{b} + B_\psi B_\varphi \mathbf{x} \,,$$

where

$$\mathbf{b} = \begin{bmatrix} a \\ b \\ c \end{bmatrix}, \quad B = B_\psi B_\varphi = \begin{bmatrix} \cos\varphi & -\sin\varphi & 0 \\ \cos\psi\sin\varphi & \cos\psi\cos\varphi & -\sin\psi \\ \sin\psi\sin\varphi & \sin\psi\cos\varphi & \cos\psi \end{bmatrix}.$$

On composing the motion with the elevation which sets $\eta = 0$ one obtains
the map $\mathbf{y} = \mathbf{b}^* + B^*\mathbf{x}$ where

$$\mathbf{b}^* = \begin{bmatrix} a \\ c \end{bmatrix}, \qquad B^* = \begin{bmatrix} \cos\varphi & -\sin\varphi & 0 \\ \sin\psi\sin\varphi & \sin\psi\cos\varphi & \cos\psi \end{bmatrix}.$$

This projection can be scaled so as to produce the final image.

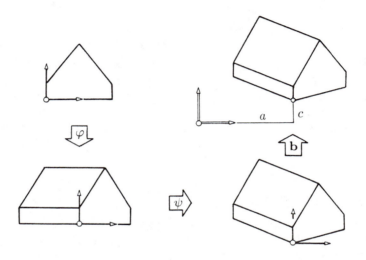

Figure 5.2: Composing motions.

A 3rd rotation around the 3-axis by φ^* may be performed before a translation, see Figure 5.3. This gives $\mathbf{y} = \mathbf{b} + B_{\varphi^*}B_{\psi}B_{\varphi}\mathbf{x}$.

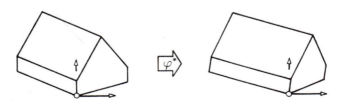

Figure 5.3: The third rotation.

5.3 Euler Angles

The three angles φ, ψ, φ^* are called **Euler angles**. However, because of historical reasons, Euler angles refer to rotations around the axes of the moved system and therefore appear in reverse order. These rotations are illustrated in Figure 5.4 by a horizontal circular disk which is

    1st rotated around the 3-axis by the angle φ^*,
    2nd rotated around the moved 1-axis by the angle ψ, and
    3rd rotated around the moved 3-axis by the angle φ.

Any object to be moved may be thought of as being fixed on and moved with the disk. In particular, the initial system with some point $\mathbf{x}$ is moved together with the disk via two intermediate steps into a final system. The final position of the point $\mathbf{x}$ has the same coordinates $\mathbf{x}$ with respect to the final system, the coordinates $B_\varphi \mathbf{x}$ with respect to the second intermidiate, the coordinates $B_\psi B_\varphi \mathbf{x}$ with respect to the first intermediate and the coordinates $B_{\varphi^*} B_\psi B_\varphi \mathbf{x}$ with respect to the initial system. Thus, the total motion with a subsequent translation is the same as before and given by

$$\mathbf{y} = \mathbf{b} + B_{\varphi^*} B_\psi B_\varphi \mathbf{x} \ .$$

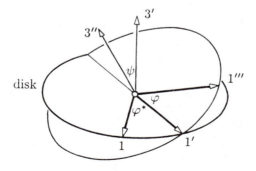

Figure 5.4: The three Euler angles.

5.4 Coordinate Extension

Successive motions will complicate the above notation, e.g., one may have

$$\mathbf{y} = \mathbf{b} + B\mathbf{x} \qquad \text{and}$$
$$\mathbf{z} = \mathbf{c} + C\mathbf{y} = [\mathbf{c} + C\mathbf{b}] + CB\mathbf{x} \ .$$

The use of 4×4 matrices can simplify the notation. Namely, the equation $\mathbf{y} = \mathbf{b} + B\mathbf{x}$ can be extended to

$$\begin{bmatrix} \mathbf{y} \\ 1 \end{bmatrix} = \left[\begin{array}{c|c} B & \mathbf{b} \\ \hline \mathbf{o}^t & 1 \end{array} \right] \begin{bmatrix} \mathbf{x} \\ 1 \end{bmatrix} ,$$

where $\mathbf{o}^t = [0\ 0\ 0]$. Hollow letters will be used to denote such an extended system, e.g., here one has $\mathbb{y} = \mathbb{B}\mathbb{x}$. The equation of the composite motion can be found via matrix multiplication, $\mathbb{z} = \mathbb{C}\mathbb{y} = \mathbb{C}\mathbb{B}\mathbb{x}$. Moreover, the associated motion of vectors $\triangle\mathbf{y} = B \triangle\mathbf{x}$ can be written using the same augmented matrix $\mathbb{B}$, i.e.,

$$\triangle\mathbb{y} = \mathbb{B}\triangle\mathbb{x} \ .$$

Augmented matrices are very popular in computer graphics to describe motions and central perspectives. The coordinate added to a point or a vector is called its **extension** and is denoted by e, where

$$e = \begin{cases} 1 \\ 0 \end{cases} \quad \text{if} \quad \begin{bmatrix} \mathbf{x} \\ e \end{bmatrix} \text{ represents a } \begin{cases} \text{point} \\ \text{vector}. \end{cases}$$

This notation reflects the facts that the difference between two points is a vector and that any multiple of a vector added to a point gives a point, but a point cannot be added to a point.

Remark 1: With extended coordinates, the coordinate system with respect to itself is represented by the four columns of the 4×4 identity matrix.

5.5 Notes and Problems

1 In Section 5.1, one can allow $\det B = -1$ in order to include **reflections**.

2 Every Euclidean motion in 3-space can be represented by a composition of a single rotation around a suitable axis in space and one translation.

3 A suitable translation should be used to center an object at the origin before rotating it.

4 There is no difference between a rotation around the 3-axis by φ and a bird's-eye view rotating around the object by $-\varphi$.

5 In a "bird's eye view" of an object sitting on the plane, the image of the z-axis should be parallel to the ζ-axis, which means that $\varphi^* = 0$.

6 The inverses of B_φ and B_ψ are $B_\varphi^{\mathrm{t}} = B_{-\varphi}$ and $B_{-\psi}^{\mathrm{t}} = B_{-\psi}$, respectively.

7 The inverse of $B = B_\psi B_\varphi$ is $B^{\mathrm{t}} = B_{-\varphi} B_{-\psi}$.

8 A "bird's eye view" from a point represented in spherical coordinates by

$$x_0 = R\cos\psi\cos\varphi \ , \quad y_0 = R\cos\psi\sin\varphi \ , \quad z_0 = R\sin\varphi$$

is given by $\mathbf{y} = \mathbf{b} + B_{-\varphi} B_{-\psi} \mathbf{x}$ for some $\mathbf{b}$.

6 Perspective Drawings

Parallel projections are not only simple to construct, but they also allow for simple reconstructions of common objects. However, if one does not need to take measurements from a projection, it may make sense to take a perspective image, as made by a camera, because of aesthetic reasons or to achieve certain realistic effects, e.g., in computer animation. The first perspective drawings were made about 550 years ago by Italian artists and architects such as Paolo Ucello (1397–1475), who drew the well-known chalice, and Leo Battista Alberti (1404–1472), who drew a true perspective view of Venice.

Literature: Hearn, Morehead, Penna·Patterson, Rehbock

6.1 Homogeneous Coordinates

Homogeneous coordinates are a powerful tool for calculating perspective projections. Let $\mathbf{x}$ be a point given by its affine coordinates x, y, and z. One sets

$$x = \frac{x_1}{x_0} , \qquad y = \frac{x_2}{x_0} , \qquad z = \frac{x_3}{x_0} ,$$

where $x_1 = x_0 x$, etc., for some $x_0 \neq 0$. The x_1, x_2, x_3, x_0 are called **homogeneous coordinates** of $\mathbf{x}$. They have the following two properties:

The coordinates x_i and ϱx_i, $\varrho \neq 0$, define the same point.

If x_0 goes to zero while x_1, x_2, x_3 remain constant, $\mathbf{x} = [x \ y \ z]^t$ goes to infinity. The coordinates x_1, x_2, x_3, x_0 are said to represent an **ideal point** or a vector if $x_0 = 0$.

Homogeneous coordinates can be obtained from the extended coordinates in Section 5.4 by multiplication with a "weight" $w \neq 0$, i.e., by setting

$$\begin{bmatrix} \mathbf{x}w \\ ew \end{bmatrix} = \begin{bmatrix} \mathbf{x} \\ e \end{bmatrix} w,$$

where $e = 1$ or $e = 0$. Homogeneous coordinates can be **inhomogeneized** simply by a division, i.e., $\mathbf{x} = \frac{\mathbf{x}w}{w}$.

Remark 1: There is essentially no difference between the solutions of the two equations $x_1 = 0$ and $x_0 = 0$. Hence the solution of $x_0 = 0$ can be viewed as a plane, just as $x_1 = 0$ represents a plane. It should be mentioned that these planes are projective planes as explained in Section 20.3.

6.2 Central Projection

The computation of a central projection is straightforward within a carefully chosen coordinate frame. Let the camera lens, i.e., the **eye**, be positioned at the origin of the Cartesian object system, and let the image system be parallel to the x- and z-axes such that its origin is at $[0 \ \delta \ 0]^t$, as illustrated in Figure 6.1.

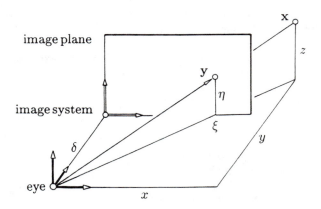

Figure 6.1: Simple central projection.

It is obvious from the intercept theorems of elementary geometry that $\xi : \delta = x : y$ and $\eta : \delta = z : y$, which means that

$$\xi = \delta \frac{x}{y}, \qquad \eta = \delta \frac{z}{y}.$$

Thus, using homogeneous coordinates, one has

$$\begin{bmatrix} \xi\omega \\ \eta\omega \\ \varepsilon\omega \end{bmatrix} = \left[\begin{array}{ccc|c} \delta & 0 & 0 & 0 \\ 0 & 0 & \delta & 0 \\ \hline 0 & 1 & 0 & 0 \end{array} \right] \begin{bmatrix} x \\ y \\ z \\ 1 \end{bmatrix}$$

where $\varepsilon = 0$ or 1. This can be written more concisely as

$$\mathsf{y}\omega = \mathbb{C}_0 \mathsf{x}.$$

Remark 2: Note that in this simple projection $\mathbf{x}$ can be scaled by any factor without changing $\mathbf{y}$. Hence one can have, for instance, x, y, z measured in yards and ξ, η measured in inches. Because of the equations for ξ and η above, δ must be measured in the same units as ξ and η.

Remark 3: The number δ is called the **eye distance**. It controls the size of the image.

6.3 Moving the Object

Any perspective projection can be obtained by first moving the object, then projecting it, and finally translating the image. Figure 6.2 illustrates this procedure with an example. (Only the elevations are shown.)

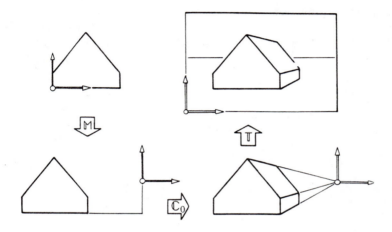

Figure 6.2: Moving, projection, and translating.

Let $\mathbb{M}$ be the 4×4 matrix of the object motion, $\mathbb{C}_0$ the 3×4 matrix of the projection, and $\mathbb{T}$ the 3×3 matrix of the image translation. Then one obtains for the example

$$\mathbb{M} = \left[\begin{array}{ccc|c} 1 & 0 & 0 & a \\ 0 & 1 & 0 & b \\ 0 & 0 & 1 & c \\ \hline 0 & 0 & 0 & 1 \end{array}\right], \quad \mathbb{C}_0 = \left[\begin{array}{ccc|c} \delta & 0 & 0 & 0 \\ 0 & 0 & \delta & 0 \\ 0 & 1 & 0 & 0 \end{array}\right], \quad \mathbb{T} = \left[\begin{array}{cc|c} 1 & 0 & \alpha \\ 0 & 1 & \beta \\ 0 & 0 & 1 \end{array}\right],$$

i.e., the total map is given by the augmented matrix

$$\mathbb{C} = \mathbb{T}\mathbb{C}_0\mathbb{M} = \left[\begin{array}{ccc|c} \delta & \alpha & 0 & \alpha b + \delta a \\ 0 & \beta & \delta & \beta b + \delta c \\ \hline 0 & 1 & 0 & b \end{array}\right].$$

In Figure 6.2 the values of a and c are negative when measured in the object system, but one requires b to be positive in order to keep the image plane between the eye and the object. More complicated motions can be handled analogously, cf. Section 7.4.

6.4 Vanishing Points

Let $\mathbf{x}$ denote a point varying on a straight line $\mathcal{L}$. If $\mathbf{x}$ is at infinity, the corresponding projecting ray is parallel to $\mathcal{L}$; it intersects the image plane at a point $\mathbf{v}$ called the **vanishing point** of $\mathcal{L}$, as illustrated in Figure 6.3.

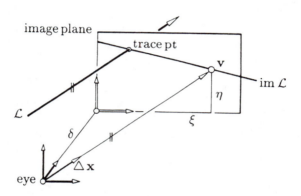

Figure 6.3: Vanishing point of a line.

Let $\Delta\mathbf{x} = [\Delta x \quad \Delta y \quad \Delta z]^{t}$ be the direction of $\mathcal{L}$. Then, in the setting of Section 6.2 one has $\xi : \delta = \Delta x : \Delta y$ and $\eta : \delta = \Delta z : \Delta y$, which can be written as $\mathsf{v}\omega = \mathbb{C}_{0}\Delta\varkappa$. If this projection was composed with a motion and a translation as in Section 6.3, one obtains instead

$$\mathsf{v}\omega = \mathbb{C}\Delta\varkappa .$$

Vanishing points have some interesting properties:

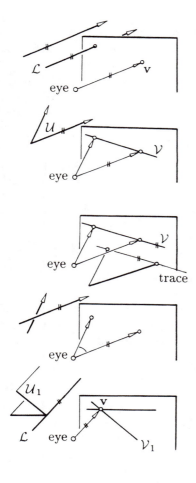

Parallel lines have the same vanishing point **v**.

The vanishing points of all lines which are parallel to a plane lie on a straight line, the **vanishing line** $\mathcal{V}$ of the plane. In particular, the vanishing points of all horizontal lines form the **horizon** $\mathcal{H}$.

The intersection of a plane with the image plane, the so-called **trace**, is parallel to the vanishing line of the plane.

The vanishing points of two straight lines are seen from the eye under the same **angle** the lines form in space.

The vanishing lines of **two planes** intersect at the vanishing point of the intersection of the planes.

Moreover, the vanishing point of a line parallel to the image plane lies at infinity. With respect to the special system in Section 6.2, the line $\mathcal{L}$ is parallel to the image plane if $\triangle y = 0$. Moreover, all points of the plane given by $y = 0$ with respect to this special system have images at infinity, with the exception of the eye whose image is not defined. This plane is called the **neutral plane**.

6.5 Completing a Perspective Drawing

The five rules above are quite powerful tools in drawing a perspective projection. One needs to know only a little information, such as a few points, in order to complete a perspective. Figure 6.4 shows an example. The marked points are computed by methods discussed in Chapter 7; all others are constructed by means of the above mentioned properties. Note that the trees and the house have the same height.

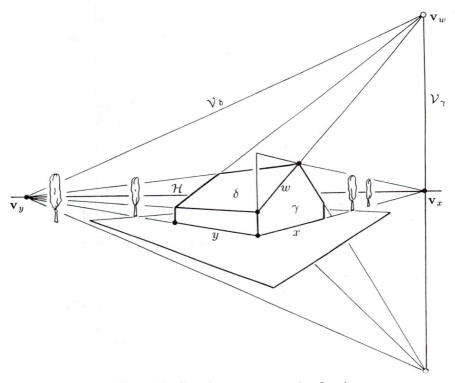

Figure 6.4: Completing a perspective drawing.

Remark 4: The perspective image of an affine scale is called a **projective scale**. It is useful in drawing a perspective image.

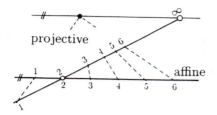

One can procure such a projective scale from an affine line in frontal position by means of a pencil of lines, as illustrated to the left.

6.6 Moving the Camera

For some applications one needs to move the camera around the object. Let the camera system after some motion be given by $\mathbf{a}; \mathbf{a}_1, \mathbf{a}_2, \mathbf{a}_3$ with respect to the object system. Then, the coordinates of a point, with respect to either the original or the final system, are related by

$$\mathbf{x} = \mathbf{a} + [\mathbf{a}_1 \ \mathbf{a}_2 \ \mathbf{a}_3] \ \mathbf{y}$$

or $\mathbf{x} = \mathbb{A}\mathbf{y}$, as illustrated in Figure 6.5.

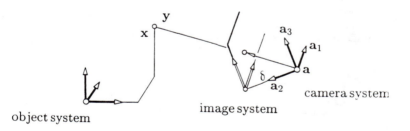

Figure 6.5: Moving the camera.

The perspective images generated by the moved camera can be computed using the inverse motion $\mathbf{y} = \mathbb{A}^{-1}\mathbf{x}$, which moves object and camera into the special position of Section 6.2, and the subsequent projection $z\varrho = \mathbb{T}\mathbb{C}_0\mathbf{y}$. Altogether, one gets for the image z of $\mathbf{x}$

$$z\sigma = \mathbb{T}\mathbb{C}_0\mathbb{A}^{-1}\mathbf{x} \ .$$

Remark 5: Since $A = [\mathbf{a_1}\ \mathbf{a_2}\ \mathbf{a_3}]$ is orthonormal, one has

$$\mathbb{A}^{-1} = \left[\begin{array}{c|c} A^t & -A^t\mathbf{a} \\ \hline \mathbf{o}^t & 1 \end{array}\right] .$$

6.7 Spatial Perspective Maps

Sometimes, for example, for hidden surface removal or shading, it is desirable to split the perspective map of Section 6.2 into a spatial perspective map and a parallel projection. This may be done by giving a temporary depth ζ to the image, while preserving spatial relationships from the object space, as illustrated in Figure 6.6.

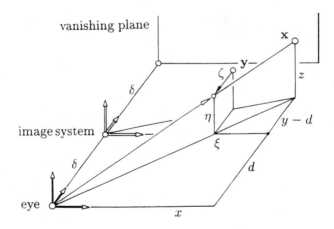

Figure 6.6: Simple spatial perspective.

Still the relations $\xi : \delta = x : y$ and $\eta : \delta = z : y$ hold, and in addition one has $\zeta : \delta = (y-d) : y$, where both d and δ denote the eye distance measured with respect to the object and image system, respectively, cf. Remark 2. Thus one has

$$\xi = \delta\,\frac{x}{y}\,,\qquad \eta = \delta\,\frac{z}{y}\,,\qquad \zeta = \delta\,\frac{y-d}{y},$$

and using homogeneous coordinates

$$
\begin{bmatrix} \xi \\ \eta \\ \zeta \\ \varepsilon \end{bmatrix} \omega = \left[\begin{array}{ccc|c} \delta & 0 & 0 & 0 \\ 0 & 0 & \delta & 0 \\ 0 & \delta & 0 & -\delta d \\ \hline 0 & 1 & 0 & 0 \end{array} \right] \begin{bmatrix} x \\ y \\ z \\ e \end{bmatrix}
$$

or more concisely $y\omega = \mathbb{R}_0 \mathbf{x}$.

The perspective presented in Section 6.2 is obtained from this spatial perspective by discarding ζ finally. Discarding ζ produces a parallel projection onto the plane $\zeta = 0$.

A simple consequence is that any affine scale in the spatial perspective corresponds to an affine scale in the perspective image and vice versa. This fact allows one to simplify certain algorithms.

Remark 6: However, a spatial perspective map changes affine ratios and angles. In particular, it does not map the normal of a plane to the normal of its spatial image.

Remark 7: The vanishing points form the so-called **vanishing plane** of the spatial perspective. Its equation is $\zeta = \delta$ since it is the image of the ideal plane $e = 0$. Note that the half space given by $y \geq d$ is compressed into the layer $0 \leq \zeta \leq \delta$.

6.8 Notes and Problems

1 In order to get a correct visual impression of an object from its perspective image, the observer has to take the **correct viewing position** used to construct the perspective. From another point one gets a more or less incorrect mental image.

2 If the viewing distance in Note 1 is greater (or smaller) than the eye distance, one gets the impression of a longer (or shorter) object.

3 Lines meeting in the neutral plane have parallel images.

4 A light source in space corresponds to a light source in a spatial perspective. In other words, the spatial perspective of a shadow is the shadow in this spatial perspective.

5 The spatial perspective introduced above can be used to shorten the depth of a scene for a stage.

6 Geometrically, one can interpret the correspondence between inhomogeneous and homogeneous coordinates as in Figure 6.7. However, it is not necessary to have this figure in mind when working with homogeneous coordinates.

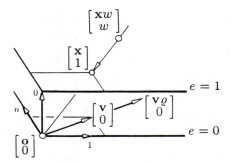

Figure 6.7: Geometric meaning of homogeneous, inhomogeneous, and extended coordinates

7 The Mapping Matrix

A general projection is composed of a motion in the object space, a projection, and a translation in the image space. Implicitly, the matrix of a perspective projection contains all this information. In particular, the matrix encompasses the camera position relative to the scene. This chapter shows how to retrieve the position of the eye and the image plane, provided that both the image and object systems are Cartesian.

Literature: Berger, Hearn, Newmann·Sproull.

7.1 Main Theorem

Consider the central projection given by

$$\begin{bmatrix} \mathbf{y} \\ \varepsilon \end{bmatrix} \omega = \mathbb{C} \begin{bmatrix} \mathbf{x} \\ e \end{bmatrix} , \qquad \text{where} \quad \mathbb{C} = \begin{bmatrix} \mathbf{a}_1\alpha_1 & \mathbf{a}_2\alpha_2 & \mathbf{a}_3\alpha_3 & \mathbf{a}\alpha \\ \varepsilon_1\alpha_1 & \varepsilon_2\alpha_2 & \varepsilon_3\alpha_3 & \alpha \end{bmatrix} ,$$

with $\varepsilon, e, \varepsilon_1, \varepsilon_2, \varepsilon_3$ being coordinate extensions as described in Section 5.4. The columns of the mapping matrix have a straightforward **geometric meaning**: As in Section 4.3, the columns of the 4×4 identity matrix represent the object system. Thus, the columns of $\mathbb{C}$ represent the image of the object system in homogeneous coordinates. By inhomogeneizing the columns, one has that

> **a** describes the image of the origin while $\mathbf{a}_i$ describes the vanishing point of the i-direction which is an ideal point if $\varepsilon_i = 0$.

Moreover, the point $\varkappa = [1\ 0\ 0\ 1]^t$ has the image

$$\begin{bmatrix} \mathbf{b}_1 \\ \varepsilon \end{bmatrix} \omega = \begin{bmatrix} \mathbf{a}_1 \\ \varepsilon_1 \end{bmatrix} \alpha_1 + \begin{bmatrix} \mathbf{a} \\ 1 \end{bmatrix} \alpha .$$

If $\mathbf{b}_1$ is not an ideal point, i.e., if $\varepsilon\omega = \alpha_1\varepsilon_1 + \alpha \neq 0$, one obtains the inhomogeneous coordinates

$$\mathbf{b}_1 = \mathbf{a}_1 \frac{\alpha_1}{\alpha_1\varepsilon_1 + \alpha} + \mathbf{a} \frac{\alpha}{\alpha_1\varepsilon_1 + \alpha} .$$

If $\varepsilon_1 = 1$, the coefficients of $\mathbf{a}_1$ and $\mathbf{a}$ sum to one, i.e., they are the barycentric coordinates of $\mathbf{b}_1$ with respect to $\mathbf{a}_1$ and $\mathbf{a}$. This means geometrically that

$\mathbf{b}_1$ divides $\mathbf{a}$ and $\mathbf{a}_1$ in the ratio $\alpha_1 : \alpha$.

If $\varepsilon_1 = 0$, one gets $\mathbf{b}_1 = \mathbf{a} + \mathbf{a}_1\alpha_1/\alpha$. Both of these cases are illustrated in Figure 7.1 for $\alpha\alpha_1 > 0$.

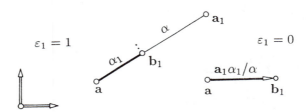

Figure 7.1: The meaning of α_1.

Note that $\alpha\alpha_1$ can be negative. Figure 7.1 would then look slightly different.

Remark 1: The image of the point $\mathbf{x} = [l\ \ 0\ \ 0]^t$ divides $\mathbf{a}$ and $\mathbf{a}_1$ in the ratio $\alpha_1 l : \alpha$, if $\varepsilon_1 = 1$. If $\varepsilon_1 = 0$, $\mathbf{x}$ has the image $\mathbf{a} + \mathbf{a}_1 l\alpha_1/\alpha$.

Remark 2: Translating the object parallel to the image plane causes $\mathbf{a}$ and $\mathbf{b}_1$ to move along parallel lines while $\mathbf{a}_1$ remains fixed, i.e., $\mathbf{b}_1$ divides $\mathbf{a}$ and $\mathbf{a}_1$ in the same ratio before and after such a translation.

7.2 Camera Data

The **neutral plane** with its ideal points consists of the points with images at infinity, i.e., all points for which $\varepsilon = 0$. Therefore

$$\varepsilon \omega = \varepsilon_1 \alpha_1 \cdot x + \varepsilon_2 \alpha_2 \cdot y + \varepsilon_3 \alpha_3 \cdot z + \alpha \cdot e = 0$$

represents the neutral plane with respect to the object system. The **camera eye** is the only point with an indefinite image. Thus, the solution of the homogeneous system

$$\mathbb{C}\varkappa = \mathbb{O}$$

represents the eye with respect to the object system. Obviously, the eye lies in the neutral plane. Note that the neutral plane is parallel to the image plane. Their normal, given by

$$\mathfrak{n} = [\varepsilon_1 \alpha_1 \quad \varepsilon_2 \alpha_2 \quad \varepsilon_3 \alpha_3 \quad 0]^t \ ,$$

is called the **principal direction**, and the vanishing point $\mathbf{h}$ of the principal direction is called the **principal point** of the image plane. Inhomogeneizing $\mathfrak{h} = \mathbb{C}\mathfrak{n}$ gives

$$\mathbf{h} = \mathbf{a}_1 \lambda_1 + \mathbf{a}_2 \lambda_2 + \mathbf{a}_3 \lambda_3 \ ,$$

where $\lambda_i = \varepsilon_i \alpha_i^2 / \mathbf{n}^t \mathbf{n}$. One can check that the λ_i sum to one. Therefore they are the barycentric coordinates of $\mathbf{h}$ with respect to $\mathbf{a}_1, \mathbf{a}_2, \mathbf{a}_3$.

Finally, let $\triangle\mathbf{x}$ be some direction which forms an angle of $45°$ with the principal ray. The distance between the vanishing point of $\triangle\mathbf{x}$ and $\mathbf{h}$ is equal to the **eye distance** δ.

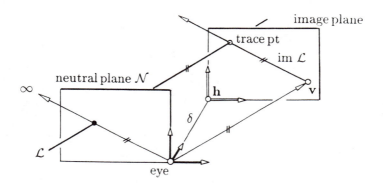

Figure 7.2: Neutral plane $\mathcal{N}$ and principal point $\mathbf{h}$.

Note that $\alpha = 0$ implies that the origin of the object system coincides with the eye.

7.3 The Spatial Perspective

Using methods similar to those in Section 6.6, one can obtain a spatial perspective from $\mathbb{C}$ by adding a further row to $\mathbb{C}$. This row corresponds to an additional coordinate ζ of the image. For instance, let

$$\zeta\omega = \varrho\,[\varepsilon_1\alpha_1 \;\; \varepsilon_2\alpha_2 \;\; \varepsilon_3\alpha_3 \;\; 0] \begin{bmatrix} \mathbf{x} \\ e \end{bmatrix} = \varrho\mathbf{n}^t\mathbf{x}$$

for some $\varrho \neq 0$. Obviously, this extended image represents a spatial perspective which reduces to the initial plane image if one sets $\zeta = 0$. On comparing the expressions for ζ and ε one finds

$$\zeta\omega = \varrho\mathbf{n}^t\mathbf{x} = \varrho(\varepsilon\omega - \alpha e)$$

which reduces to $\zeta = \varrho\varepsilon$ if $e = 0$. The geometric meaning of these expressions is that the vanishing points form a plane, the vanishing plane $\zeta = \varrho\varepsilon$. This plane is parallel to the image plane $\zeta = 0$.

Remark 3: Let $\mathbf{u}^t\mathbf{x} + u_0 = 0$ be the equation of a plane in the object space, and let $\mathbf{v}^t\mathbf{y} + v_0 = 0$ be the equation of its image in the image space. Then the planes are related by the equation

$$\sigma[\mathbf{u}^t, u_0] = [\mathbf{v}^t, v_0] \begin{bmatrix} \mathbf{a}_1\alpha_1 & \mathbf{a}_2\alpha_2 & \mathbf{a}_3\alpha_3 & \mathbf{a}\alpha \\ \varrho\varepsilon_1\alpha_1 & \varrho\varepsilon_2\alpha_2 & \varrho\varepsilon_3\alpha_3 & 0 \\ \varepsilon_1\alpha_1 & \varepsilon_2\alpha_2 & \varepsilon_3\alpha_3 & \alpha \end{bmatrix},$$

where σ is some scalar. This correspondence is useful in the computation of light effects, etc.

7.4 Vanishing Points of the System

The three directions of the (Cartesian) object system are mapped into finite or infinite vanishing points. One can classify a general central projection by the number of these vanishing points which are finite. For the sake of simplicity, $\alpha = 1$ is assumed.

Parallel projection: If all three ε_i vanish, the mapping matrix $\mathbb{C}$ represents a parallel projection, as considered in Chapter 4.

One-point perspective: If there is only one finite vanishing point, say $\mathbf{a}_2$, $\mathbb{C}$ is of the form

$$\mathbb{C} = \begin{bmatrix} \mathbf{a}_1 & \mathbf{a}_2\alpha_2 & \mathbf{a}_3 & \mathbf{a} \\ 0 & \alpha_2 & 0 & 1 \end{bmatrix}.$$

Figure 7.3 illustrates the meaning of the $\mathbf{a}$'s and α_2 for $\alpha_2 > 0$. From Section 7.2, it follows that

$$\mathbf{h} = \mathbf{a}_2 \quad \text{and} \quad \delta = |\mathbf{a}_1|/|\alpha_2| = |\mathbf{a}_3|/|\alpha_2| .$$

Because of this, and since the 1- and 3-directions of the object system are orthogonal and parallel to the image plane, one has

$$|\mathbf{a}_1| = |\mathbf{a}_3| \quad \text{and} \quad \mathbf{a}_1 \perp \mathbf{a}_3 .$$

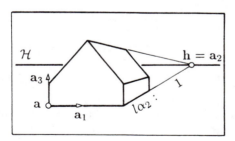

Figure 7.3: One-point perspective, $\alpha_2 > 0$.

Two-point perspective: If there are exactly two finite vanishing points, say $\mathbf{a}_1$ and $\mathbf{a}_2$, $\mathbb{C}$ is of the form

$$\mathbb{C} = \begin{bmatrix} \mathbf{a}_1\alpha_1 & \mathbf{a}_2\alpha_2 & \mathbf{a}_3 & \mathbf{a} \\ \alpha_1 & \alpha_2 & 0 & 1 \end{bmatrix}.$$

Figure 7.4 illustrates the meaning of the $\mathbf{a}$'s and α's. From Section 7.2, it follows immediately that

$$\mathbf{h} = \mathbf{a}_1\lambda_1 + \mathbf{a}_2\lambda_2, \qquad \text{where} \quad \lambda_i = \frac{\alpha_i^2}{\alpha_1^2 + \alpha_2^2} \, ,$$

and after some algebraic manipulations

$$\delta = \frac{|\mathbf{a}_3|}{\sqrt{\alpha_1^2 + \alpha_2^2}} = \frac{|\alpha_1\alpha_2|}{\alpha_1^2 + \alpha_2^2} |\mathbf{a}_2 - \mathbf{a}_1| \, .$$

As above, the $\mathbf{a}$'s and α's are not independent. Namely, one has

$$\mathbf{a}_3 \perp \mathbf{a}_2 - \mathbf{a}_1 \quad \text{and} \quad |\mathbf{a}_3|\sqrt{\alpha_1^2 + \alpha_2^2} = |\alpha_1\alpha_2| \cdot |\mathbf{a}_2 - \mathbf{a}_1| \, .$$

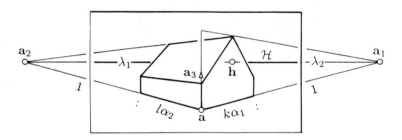

Figure 7.4: Two-point perspective, $\alpha_1 > 0$, $\alpha_2 > 0$.

Three-point perspective: If there are three finite vanishing points, $\mathbb{C}$ is of the most general form

$$\mathbb{C} = \begin{bmatrix} \mathbf{a}_1\alpha_1 & \mathbf{a}_2\alpha_2 & \mathbf{a}_3\alpha_3 & \mathbf{a} \\ \alpha_1 & \alpha_2 & \alpha_3 & 1 \end{bmatrix} \, .$$

Figure 7.5 illustrates the meaning of the $\mathbf{a}$'s and α's. Let $\mathbf{e}$ denote the eye. The plane $\mathbf{e}\,\mathbf{h}\,\mathbf{a}_3$ is perpendicular to the planes $\mathbf{a}_1\mathbf{a}_2\mathbf{e}$ and $\mathbf{a}_1\mathbf{a}_2\mathbf{a}_3$ and consequently also to the line $\mathbf{a}_1\mathbf{a}_2$. Therefore the lines $\mathbf{h}\,\mathbf{a}_3$ and $\mathbf{a}_1\mathbf{a}_2$

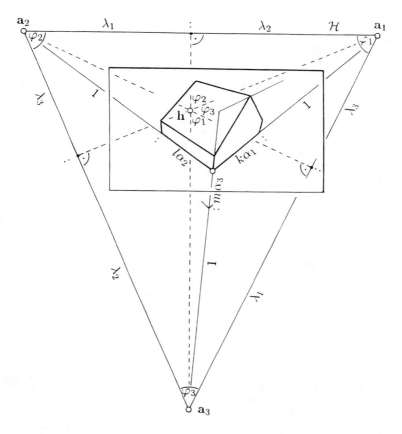

Figure 7.5: Three-point perspective, $\alpha_i > 0$.

are perpendicular. Then because of symmetry, $\mathbf{h}$ is the orthocenter of the triangle $\mathbf{a}_1\mathbf{a}_2\mathbf{a}_3$, i.e., one has

$$\lambda_1 : \lambda_2 : \lambda_3 = \tan\varphi_1 : \tan\varphi_2 : \tan\varphi_3 \;,$$

where the λ_i are the barycentric coordinates of $\mathbf{h}$ with respect to the $\mathbf{a}_i$, and the φ_i are as defined by Figure 7.5. From Section 7.2, it follows that the λ_i are positive and that

$$\alpha_1^2 : \alpha_2^2 : \alpha_3^2 = \lambda_1 : \lambda_2 : \lambda_3 \;,$$

i.e., the α_i are not independent. Finally, one gets after some calculations

$$\delta = \frac{\sqrt{\lambda_1 \lambda_2}}{\sqrt{\lambda_1 + \lambda_2}} |\mathbf{a}_2 - \mathbf{a}_1| .$$

Example 1: An example of a one-point perspective was considered in Section 6.3, where

$$\mathbb{C} = \mathbb{T}\mathbb{C}_0 \mathbb{M} = \left[\begin{array}{ccc|c} \delta & \alpha & 0 & \alpha b + \delta a \\ 0 & \beta & \delta & \beta b + \delta c \\ \hline 0 & 1 & 0 & b \end{array} \right] .$$

The camera is positioned at $\varkappa_1 = [-a \ -b \ -c \ \ 1]^t$ and looks in the direction of $\mathbf{n}_1 = [0 \ 1 \ 0]^t$.

Example 2: One gets a two-point perspective if the projection of Example 1 is composed with the motion of Example 2 in Section 5.1. The matrix

$$\mathbb{C} = \mathbb{T}\mathbb{C}_0 \mathbb{M} \mathbb{B}_\varphi = \left[\begin{array}{ccc|c} \alpha s_\varphi + \delta c_\varphi & \alpha c_\varphi - \delta s_\varphi & 0 & \alpha b + \delta a \\ \beta s_\varphi & \beta c_\varphi & \delta & \beta b + \delta c \\ \hline s_\varphi & c_\varphi & 0 & b \end{array} \right]$$

is the mapping matrix where $s_\varphi = \sin\varphi$ and $c_\varphi = \cos\varphi$. The camera is positioned at $\varkappa_2 = \mathbb{B}_\varphi^t \varkappa_1$ and looks in the direction of $\mathbf{n}_2 = [s_\varphi \ c_\varphi \ 0]^t$.

Example 3: One gets a three-point perspective if the projection of Example 2 is composed with the rotation of Example 3 in Section 5.1. Let c and s be abbreviations for cosine and sine; the new mapping matrix is

$$\mathbb{C} = \mathbb{T}\mathbb{C}_0 \mathbb{M} \mathbb{B}_\psi \mathbb{B}_\varphi =$$
$$= \left[\begin{array}{ccc|c} \delta c_\varphi + \alpha c_\psi s_\varphi & -\delta s_\varphi + \alpha c_\psi c_\varphi & -\alpha s_\psi & \delta a + \alpha b \\ \delta s_\psi s_\varphi + \beta c_\psi s_\varphi & \delta s_\psi c_\varphi + \beta c_\psi c_\varphi & \delta c_\psi - \beta s_\psi & \delta c + \beta b \\ \hline c_\psi s_\varphi & c_\psi c_\varphi & -s_\psi & b \end{array} \right] .$$

The camera is positioned at $\varkappa_3 = \mathbb{B}_\varphi^t \mathbb{B}_\psi^t \varkappa_1$ and looks in the direction of $\mathbf{n}_3 = [c_\psi s_\varphi \ c_\psi s_\varphi \ -s_\psi]^t$.

7.5 Stereo Pairs

In every planar perspective drawing of the 3-space, one dimension is lost. A stereo pair, i.e., a pair of perspective images, as illustrated in Figure 7.6, can compensate for this loss.

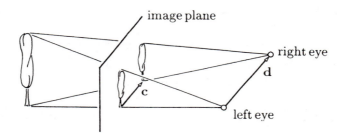

Figure 7.6: The concept of stereo pairs.

It is assumed that both perspective projections have the same image plane and the same neutral plane. The subscripts L and R are used to indicate the position of the eye. Let

$$\mathbb{C}_L = \begin{bmatrix} \mathbf{a}_1\alpha_1 & \mathbf{a}_2\alpha_2 & \mathbf{a}_3\alpha_3 & \mathbf{a}_L \\ \varepsilon_1\alpha_1 & \varepsilon_2\alpha_2 & \varepsilon_3\alpha_3 & 1 \end{bmatrix}$$

be the matrix of the left projection. Let $\mathbf{d}$ denote the vector from the left to the right eye with respect to the image system, and let $\mathbf{c}$ be the vector from $\mathbf{a}_L$ to $\mathbf{a}_R$. Then $\mathbf{c}$ and $\mathbf{d}$ are parallel. From the main theorem in Section 7.1 with Remark 2 it follows that

$$\mathbb{C}_R = \begin{bmatrix} [\mathbf{a}_1 + \mathbf{d}\varepsilon_1]\alpha_1 & [\mathbf{a}_2 + \mathbf{d}\varepsilon_2]\alpha_2 & [\mathbf{a}_3 + \mathbf{d}\varepsilon_3]\alpha_3 & \mathbf{a}_L + \mathbf{c} \\ \varepsilon_1\alpha_1 & \varepsilon_2\alpha_2 & \varepsilon_3\alpha_3 & 1 \end{bmatrix} .$$

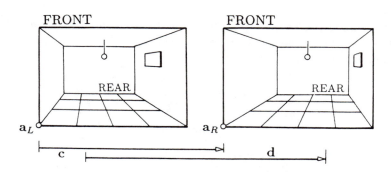

Figure 7.7: Stereo pair, example.

Example 4: Figure 7.7 shows an example with simplified entries,

$$
\mathbb{C}_L = \begin{bmatrix} f & a\beta & 0 & 0 \\ 0 & b\beta & f & 0 \\ 0 & 1\beta & 0 & 1 \end{bmatrix}, \qquad
\mathbb{C}_R = \begin{bmatrix} f & (a+d)\beta & 0 & c \\ 0 & b\beta & f & 0 \\ 0 & 1\beta & 0 & 1 \end{bmatrix}.
$$

One can try to see a 3D-object with both eyes being at a distance of 1 foot. An application is given in Section 8.3.

Remark 4: Let γ be such that $\mathbf{c} = \mathbf{d}\gamma$. If the origin of the object system lies behind, on, or in front of the image plane, but not behind the eyes, one has $0 < \gamma < 1$, $\gamma = 0$, or $\gamma < 0$, respectively.

7.6 Notes and Problems

1 The image $\mathbf{y}$ of a point $\mathbf{x}$ is an affine combination

$$
\mathbf{y} = \frac{\mathbf{a}_1\alpha_1 x + \cdots + \mathbf{a}_3\alpha_3 z + \mathbf{a}\alpha}{\varepsilon_1\alpha_1 x + \cdots + \varepsilon_3\alpha_3 z + \alpha}
$$

of the four points (or vectors) $\mathbf{a}_1, \mathbf{a}_2, \mathbf{a}_3$ and $\mathbf{a}$. Note that some of the ε_i may be zero and that some of the α_i may be negative.

2 The image $\mathbf{y}$ of a point $\mathbf{x} = [x \quad y \quad 0]^t$ of the xy-plane is a barycentric combination

$$\mathbf{y} = \frac{\mathbf{a}_1 \alpha_1 x + \mathbf{a}_2 \alpha_2 y + \mathbf{a}\alpha}{\varepsilon_1 \alpha_1 x + \varepsilon_2 \alpha_2 y + \alpha}$$

of the three points $\mathbf{a}_1, \mathbf{a}_2$ and $\mathbf{a}$.

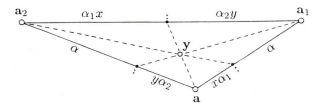

Figure 7.8: Barycentric coordinates of $\mathbf{y}$.

3 Let $\mathbf{b}_x$ be the image of $[x \quad 0 \quad 0]^t$. If $\alpha_1 x < 0 < \alpha$, then $\mathbf{a}$ divides the line segment $\mathbf{b}_x, \mathbf{a}_1$ in the ratio $-\alpha_1 x : \alpha + \alpha_1 x$, as illustrated in Figure 7.9.

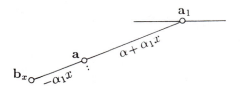

Figure 7.9: The case $0 < -\alpha_1 x < \alpha$.

4 In general, the perspective image of a perspective drawing of some object is not a perspective projection of this object.

5 Let $\mathbf{y}_R$ and $\mathbf{y}_L$ be the two stereoscopic images of the (possibly ideal) point $[\mathbf{x}^t \; e]^t$. Their difference satisfies

$$\mathbf{y}_R - \mathbf{y}_L = \mathbf{d} - (\mathbf{d} - \mathbf{c})\frac{e}{\omega},$$

where $\omega = \varepsilon_1\alpha_1 x + \varepsilon_2\alpha_2 y + \varepsilon_3\alpha_3 z + 1$. This formula can be used to simplify the computation of $\mathbf{y}_R$ and $\mathbf{y}_L$.

8 Reconstruction

Since one dimension is lost in the projection of a 3D-object into the plane, difficulties arise when one tries to retrieve the exact data of the 3D-object. Building on the methods of the previous chapter, this chapter develops strategies which can facilitate the reconstruction of object data from one or more perspective images.

Literature: Hohenberg, Penna·Patterson, Rehbock

8.1 Knowing the Object

If a photograph shows the image of a cuboid, one can read the mapping matrix $\mathbb{C}$ directly from the photograph as demonstrated in Chapter 7. In this and the following two sections, it is assumed that $\mathbb{C}$ is already determined. In Section 8.4 a general method for computing $\mathbb{C}$ is given. Let $\mathbf{y}_0$ be the known image of some unknown point $\mathbf{x} = [x \quad y \quad z]^t$. Using the notation of Chapter 7 one obtains

$$\mathbb{C}\begin{bmatrix} \mathbf{x} \\ 1 \end{bmatrix} = \begin{bmatrix} a_1\alpha_1 & a_2\alpha_2 & a_3\alpha_3 & a \\ e_1\alpha_1 & e_2\alpha_2 & e_3\alpha_3 & 1 \end{bmatrix}\begin{bmatrix} \mathbf{x} \\ 1 \end{bmatrix} = \begin{bmatrix} \mathbf{y}_0 \\ 1 \end{bmatrix}\omega \ ,$$

which represents an underdetermined system with three linear equations and four unknowns x, y, z, and ω. Further knowledge about the object is needed to determine $\mathbf{x}$.

For example, if $\mathbf{x}$ lies in a known plane, the equation of the plane

$$ax + by + cz + d = 0$$

establishes a fourth linear equation. In general these four equations are sufficient to compute $\mathbf{x}$ and the non-interesting ω. Figure 8.1 illustrates the special case where $c = 0$ and $d = -1$.

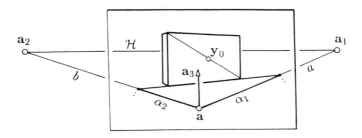

Figure 8.1: Simple case of a further condition.

8.2 Straight Lines in the Image Plane

Object data can also be reconstructed with the aid of straight lines in the image plane. Let $\mathbf{v}^t\mathbf{y} = \mathbf{v}^t\mathbf{y} + v_0 = 0$ represent some line $\mathcal{V}$ in the image plane. Multiplying $\mathbf{v}^t\mathbf{y}$ by $\omega \neq 0$ and substituting $\mathbb{C}\mathbf{x}$ for $\mathbf{y}\omega$ one gets the equation of some plane $\mathcal{U}$ in the object space

$$\mathbf{v}^t\mathbf{y}\omega = \begin{bmatrix} \mathbf{v}^t & v_0 \end{bmatrix} \begin{bmatrix} \mathbf{a}_1\alpha_1 & \mathbf{a}_2\alpha_2 & \mathbf{a}_3\alpha_3 & \mathbf{a} \\ e_1\alpha_1 & e_2\alpha_2 & e_3\alpha_3 & 1 \end{bmatrix} \begin{bmatrix} \mathbf{x} \\ 1 \end{bmatrix} = \mathbf{u}^t\mathbf{x} + u_0 = 0 .$$

Obviously $\mathcal{V}$ is the image of $\mathcal{U}$, i.e., $\mathcal{U}$ is spanned by $\mathcal{V}$ and the eye as illustrated in Figure 8.2.

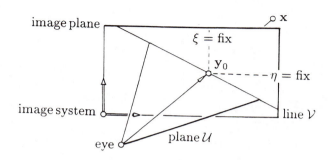

Figure 8.2: Projecting plane.

Each line $\mathcal{V}$ through some fixed point $\mathbf{y}_0$ corresponds to a plane $\mathcal{U}$ which contains the projecting ray through the eye and $\mathbf{y}_0$. The equations of two such planes form an underdetermined linear system for the corresponding object point $\mathbf{x}$. Note that three planes obtained from lines through $\mathbf{y}_0$ have linearly dependent equations. Thus, one needs a further independent condition to determine $\mathbf{x}$.

Remark 1: The line $\mathcal{V}$ is the vanishing line of all planes parallel to $\mathcal{U}$.

Remark 2: In particular, if $\mathbf{v}^t = [1\ 0\ v_0]$, $\mathcal{V}$ has the equation $\xi = -v_0$. Analogously, if $\mathbf{v}^t = [0\ 1\ v_0]$, $\mathcal{V}$ has the equation $\eta = -v_0$. These lines are used in Example 1 of the next section.

8.3 Several Images

Let $\mathbf{y}_1$, $\mathbf{y}_2$ denote the images of a point $\mathbf{x}$ in two distinct photographs with each image lying on a pair of lines as illustrated in Figure 8.3. The equations of the corresponding two pairs of planes constitute a generally overdetermined system for $\mathbf{x}$.

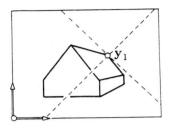 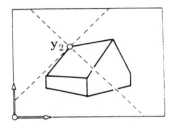

Figure 8.3: Pair of images with pairs of 45°-lines.

Each additional photograph leads to an additional pair of planes and equations, if $\mathbf{x}$ is visible. The corresponding total linear system is, in general, overdetermined and can be solved by the least squares method of Chapter 3.

Example 1: Consider the stereo pair of Example 4 in Section 7.5 and the horizontal and vertical lines through $\mathbf{y}_1 = [\xi_1, \eta]^t$ and $\mathbf{y}_2 = [\xi_2, \eta]^t$. The corresponding two pairs of planes give the equations

$$
\begin{bmatrix} 1 & 0 & -\xi_1 \\ 0 & 1 & -\eta \end{bmatrix}
\begin{bmatrix} f & a\beta & 0 & 0 \\ 0 & b\beta & f & 0 \\ 0 & \beta & 0 & 1 \end{bmatrix}
\begin{bmatrix} x \\ y \\ z \\ 1 \end{bmatrix}
= \begin{bmatrix} 0 \\ 0 \end{bmatrix} ,
$$

and

$$
\begin{bmatrix} 1 & 0 & -\xi_2 \\ 0 & 1 & -\eta \end{bmatrix}
\begin{bmatrix} f & (a+d)\beta & 0 & c \\ 0 & b\beta & f & 0 \\ 0 & \beta & 0 & 1 \end{bmatrix}
\begin{bmatrix} x \\ y \\ z \\ 1 \end{bmatrix}
= \begin{bmatrix} 0 \\ 0 \end{bmatrix} ,
$$

i.e.,

$$
\begin{aligned}
fx + \beta(a - \xi_1)y \quad\quad &= \xi_1, \\
\beta(b - \eta)y + fz &= \eta
\end{aligned}
$$

and

$$
\begin{aligned}
fx + \beta(a + d - \xi_2)y &= \xi_2 - c, \\
\beta(b - \eta)y + fz &= \eta,
\end{aligned}
$$

respectively. Note that in this example two of the four planes coincide.

Remark 3: To avoid problems with coinciding planes in the case of a horizontally moved camera, as in Example 1, the lines should not be horizontal.

Remark 4: One may add further conditions, as in Figure 8.1, in order to get an overdetermined system.

8.4 Camera Calibration

Straight lines in the image plane can also be used to determine the mapping matrix $\mathbb{C}$ from some known points and their known images. Let

$$\mathbb{p} = [p_1 \ p_2 \ p_3 \ p_0]^t$$

be the homogeneous coordinates of some point $\mathbb{p}$, and let

$$\mathbb{C} = [c_{i,j}] = \begin{bmatrix} c_{1,1} & c_{1,2} & c_{1,3} & c_{1,0} \\ c_{2,1} & c_{2,2} & c_{2,3} & c_{2,0} \\ c_{0,1} & c_{0,2} & c_{0,3} & c_{0,0} \end{bmatrix}$$

be the unknown mapping matrix. The condition that the image $\mathbb{q}$ of $\mathbb{p}$ lies on a line $\mathcal{V}$ is expressed by the equation

$$\mathbb{v}^t \mathbb{q}\omega = \mathbb{v}^t \mathbb{C}\mathbb{p} = \sum_{i,j} v_i p_j c_{i,j} = 0 \ .$$

For any given $\mathbb{v}$ and any given $\mathbb{p}$ this is one homogeneous linear equation for the twelve unknowns $c_{i,j}$. The coefficient of $c_{i,j}$ is simply $v_i p_j$. Thus one needs at least eleven lines, e.g., six points $\mathbb{p}$ and their images $\mathbb{q}$, each on a pair of lines. If more lines and points are used one can employ the least squares method in Section 3.1.

If $\mathbb{C}$ is computed, the camera position and the principal ray in the object system may be computed by the methods discussed in Section 7.2. The combined procedure is called **camera calibration**.

Remark 5: A point $\mathbb{p}$ may also be replaced by a direction where $p_0 = 0$. Then, consequently, $\mathbb{q}$ represents a vanishing point; see also Note 4.

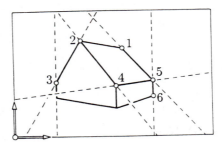

Figure 8.4: Points and lines for a camera calibration.

8.5 Notes and Problems

1 For a stereo pair, image lines parallel to **d** define only one plane in the object space, as demonstrated in Example 1.

2 In Section 8.2, a factor may be given to a projective plane $\mathbf{u}^t\mathbf{x} + u_0 = 0$ which is approximatively reciprocal to the distance between the involved point **x** and the eye so as to weight the equation and manipulate its influence.

3 When computing $\mathbb{C}$, as discussed in Section 8.4, one can set $c_{0,0} = 1$.

4 The immediate construction of $\mathbb{C}$ by the main theorem Section 7.1 may be viewed as a special example of camera calibration.

5 Since $\mathbb{C}$ is homogeneous it depends only on 11 parameters. However, these parameters are not independent. They must satisfy two constraints as described in Section 7.4. One can check also directly that the camera position relative to the image plane has 9 degrees of freedom.

PART THREE

Affine Geometry

Transformations which map lines into lines and also preserve parallelism and ratios are called affine due to Leonid Euler (1707–1783). The parallel projections and scalings discussed in Chapter 4 are examples of affine maps.

In his inaugural address at the University of Erlangen in 1872, the famous **Erlangener Programm**, Felix Klein (1849–1925) distinguished the different geometries by the properties and theorems which remain valid under certain groups of transformations. Affine geometry consists of all propositions left invariant under affine maps.

Many concepts, tools, and objects in geometric design, including the notions of smoothness, tangents, and control points, linear interpolation, and quadrics, belong to or possess an affine structure. The corresponding constructions, e.g., of points and tangents, subdivision algorithms, etc., are often also invariant under affine maps. Such invariance makes these constructions very valuable in practical computer applications.

9 Affine Space

An **affine space** $\mathcal{A}$ is a point space associated with a linear space $\mathbf{A}$ as in Section 2.5. The dimension of $\mathcal{A}$ is defined as the dimension of $\mathbf{A}$. The solution of an inhomogeneous linear system is an abstract example of an affine space.

Literature: Berger, Greub, Schaal

9.1 Affine Coordinates

Let $\mathbf{a}$ be a fixed point of an affine space $\mathcal{A}$ and let $\mathbf{a}_1, \ldots, \mathbf{a}_n$ be a basis of the associated linear space $\mathbf{A}$. For simplicity, regard points and vectors as elements of $\mathbb{R}^d$ where $d \geq n$. Then, by the properties mentioned in Section 2.5, a point $\mathbf{p}$ of $\mathcal{A}$ has a unique representation

$$\mathbf{p} = \mathbf{a} + \mathbf{a}_1 x_1 + \cdots + \mathbf{a}_n x_n \ ,$$

or in compact form $\mathbf{p} = \mathbf{a} + A\mathbf{x}$. The $x_1, \ldots, x_n$ are called **affine coordinates** of $\mathbf{p}$ with respect to the **affine system** $\mathbf{a}; \mathbf{a}_1, \ldots, \mathbf{a}_n$. Let $\mathbf{q} = \mathbf{a} + A\mathbf{y}$ denote a second point. Then the vector $\mathbf{p} - \mathbf{q}$ has the coordinates $\mathbf{x} - \mathbf{y}$. Sometimes the notation $\mathcal{A}^n$ is used to indicate that $\mathcal{A}$ has dimension n.

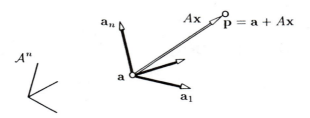

Figure 9.1: Affine system.

Every linear space has the structure of an affine space since the coordinates of a vector can be treated as the coordinates of a point and the zero vector can be interpreted as the origin.

9.2 Affine Subspaces

An affine subspace $\mathcal{B}$ of $\mathcal{A}$ is determined by a point $\mathbf{b}$ and a subspace $\mathbf{B}$ of the linear space $\mathbf{A}$. Let $\mathbf{b}$ be given as a coordinate column with respect to some affine system $\mathbf{a}; \mathbf{a}_1, \ldots, \mathbf{a}_n$ of $\mathcal{A}$. Furthermore, let $\mathbf{b}_1, \ldots, \mathbf{b}_r$ be a basis of $\mathbf{B}$ given by its coordinates with respect to the basis $\mathbf{a}_1, \ldots, \mathbf{a}_n$ of $\mathbf{A}$. Then any point $\mathbf{x} \in \mathcal{B}$ has a unique representation

$$\mathbf{x} = \mathbf{b} + \mathbf{b}_1 y_1 + \cdots + \mathbf{b}_r y_r ,$$

compactly written as $\mathbf{x} = \mathbf{b} + B\mathbf{y}$,

$$n \begin{array}{|c|} \hline \\ \mathbf{x} \\ \\ \hline \end{array} = \begin{array}{|c|} \hline \\ \mathbf{b} \\ \\ \hline \end{array} + \begin{array}{|c|} \hline \\ B \\ \\ \hline \end{array} \begin{array}{|c|} \hline \mathbf{y} \\ \hline \end{array} r \quad .$$

$$\underset{1}{} \quad \underset{1}{} \quad \underset{r}{} \quad \underset{1}{}$$

Obviously, the $y_1, \ldots, y_r$ are affine coordinates of $\mathbf{x}$ in the affine system $\mathbf{b}; \mathbf{b}_1, \ldots, \mathbf{b}_r$ of $\mathcal{B}$. The linear subspace $\mathbf{B}$ is called the **total direction** or simply the **direction** of $\mathcal{B}$. One has $dim\,\mathcal{B} = dim\,\mathbf{B} = r$.

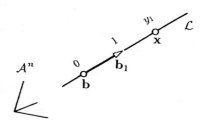

Example 1: An **affine line** $\mathcal{L}$ is represented by

$$\mathbf{x} = \mathbf{b} + \mathbf{b}_1 y_1 \ .$$

One has $dim\,\mathcal{L} = 1$. The affine coordinates y_1 of the points $\mathbf{x}$ are said to form an **affine scale** on $\mathcal{L}$.

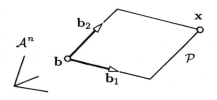

Example 2: An **affine plane** $\mathcal{P}$ is represented by

$$\mathbf{x} = \mathbf{b} + \mathbf{b}_1 y_1 + \mathbf{b}_2 y_2 \ .$$

One has $dim\,\mathcal{P} = 2$.

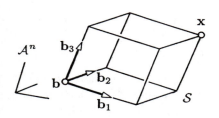

Example 3: An **affine space** $\mathcal{S}$ is represented by

$$\mathbf{x} = \mathbf{b} + \mathbf{b}_1 y_1 + \mathbf{b}_2 y_2 + \mathbf{b}_3 y_3 .$$

One has $dim\,\mathcal{S} = 3$.

Example 4: Let $r = n$. Then the subspace $\mathcal{B} \subset \mathcal{A}$ coincides with $\mathcal{A}$ and the $y_1, \ldots, y_n$ are **new affine coordinates** with respect to the **new affine system** $\mathbf{b}; \mathbf{b}_1, \ldots, \mathbf{b}_n$.

Remark 1: A subspace $\mathcal{C}$ of a subspace $\mathcal{B}$ can be represented with respect to an affine system of $\mathcal{B}$. If

$$\mathbf{y} = \mathbf{c} + C\mathbf{z}$$

is such a representation of $\mathcal{C}$, then $\mathbf{x} = \mathbf{b} + B\mathbf{y} = \mathbf{b} + B\mathbf{c} + BC\mathbf{z}$ describes $\mathcal{C}$ with respect to an affine system of $\mathcal{A}$. For example, a line in the affine plane $\mathcal{P}$ of Example 2 is represented by

$$\mathbf{y} = \begin{bmatrix} y_1 \\ y_2 \end{bmatrix} = \mathbf{c} + \mathbf{c}_1 z_1 \ ,$$

where z_1 forms an affine scale on $\mathcal{L}$.

Remark 2: Any affine system of a subspace $\mathcal{B}$ of $\mathcal{A}$ can be extended to an affine system of $\mathcal{A}$.

9.3 Hyperplanes

An affine subspace $\mathcal{U}$ of $\mathcal{A}$ with $dim\,\mathcal{U} = dim\,\mathcal{A} - 1$ is called a **hyperplane** in $\mathcal{A}$. From its **parametric representation**

$$\mathbf{x} = \mathbf{b} + \mathbf{b}_1 y_1 + \cdots + \mathbf{b}_{n-1} y_{n-1}$$

one can infer that $\mathbf{x}$ solves an inhomogeneous linear system with one equation,

$$\mathbf{u}^t \mathbf{x} + u_0 = 0 \ ,$$

see Section 1.4, Remark 3. This **equation** of the hyperplane is uniquely defined up to a factor. Another (converse) consequence of the Gauss-Jordan algorithm is that the solution of a single equation forms a hyperplane.

More generally one has that an affine subspace $\mathcal{B}$ of dimension r defined by

$$\mathbf{x} = \mathbf{b} + B\mathbf{y}$$

can be regarded as the solution of an inhomogeneous system,

$$A\mathbf{x} = \mathbf{a} \ ,$$

with $n - r$ independent equations. Here, $\mathcal{B}$ is the intersection of $n - r$ independent hyperplanes. The (explicit) parameter representation of $\mathcal{B}$ can be converted into the **implicit representation** $A\mathbf{x} = \mathbf{a}$ by means of the Gauss-Jordan algorithm and vice versa.

Remark 3: The i-axis of the underlying affine system intersects the hyperplane $\mathbf{u}^t\mathbf{x} + u_0 = 0$ at the abscissa $x_i = u_0/u_i$. Hence, if $u_0 \neq 0$, the equation can be written as

$$\frac{x_1}{a_1} + \cdots + \frac{x_n}{a_n} = 1 ,$$

where $a_i = u_0/u_i$. This is called the **intercept equation** of the hyperplane.

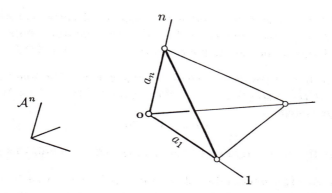

Figure 9.2: The intercepts of a plane.

Example 5: The plane given by the equation

$$1x_1 + 2x_2 + 3x_3 = 4$$

has the parameter representation

$$\mathbf{x} = \begin{bmatrix} 4 \\ 0 \\ 0 \end{bmatrix} + \begin{bmatrix} 2 & 3 \\ -1 & 0 \\ 0 & -1 \end{bmatrix} \mathbf{y}$$

and the **intercepts** $a_1 = 4$, $a_2 = 2$, $a_3 = 4/3$.

9.4 Intersection

Let $\mathcal{A}$ and $\mathcal{B}$ be two subspaces of some n-dimensional affine space given by the two following systems of linear equations:

$$A\mathbf{x} = \mathbf{a} \quad \text{and} \quad B\mathbf{x} = \mathbf{b} .$$

The **intersection** of $\mathcal{A}$ and $\mathcal{B}$, i.e., the set of all common points of $\mathcal{A}$ and $\mathcal{B}$, forms the solution of the system

$$\begin{bmatrix} A \\ B \end{bmatrix} \mathbf{x} = \begin{bmatrix} \mathbf{a} \\ \mathbf{b} \end{bmatrix} .$$

Therefore the intersection of $\mathcal{A}$ and $\mathcal{B}$ is again an affine subspace. It is written as $\mathcal{A} \sqcap \mathcal{B}$. Note that the intersection may be empty.

Often, the equations of the composite system are linearly dependent, and one can reduce their number by Gaussian eliminations. Note that the number of linearly independent equations equals $n - dim\,\mathcal{A} \sqcap \mathcal{B}$.

If $\mathcal{A}$ is given by a parameter representation $\mathbf{x} = \mathbf{c} + C\mathbf{y}$ while $\mathcal{B}$ is represented by $B\mathbf{x} = \mathbf{b}$, one can compute the intersection by a substitution. The linear system

$$BC\mathbf{y} = \mathbf{b} - B\mathbf{c}$$

represents the intersection with respect to the affine coordinates $\mathbf{y}$ of $\mathcal{A}$.

Remark 4: The intersection of $\mathcal{A} \sqcap \mathcal{B}$ can be computed in the same fashion as above if $\mathcal{A}$ and $\mathcal{B}$ are given by barycentric coordinates which are introduced in Chapter 10.

9.5   Parallel Bundles

Let $\mathbf{u}^t\mathbf{x} + u_0 = 0$ be the equation and $\mathbf{x} = \mathbf{b} + B\mathbf{y}$ be the parameter representation of some hyperplane. By the Gauss-Jordan algorithm, the direction $B\mathbf{y}$ is the solution of the homogeneous equation $\mathbf{u}^t\mathbf{x} = 0$. Hence, by varying u_0 one obtains a one-parameter family of hyperplanes with the same direction. This family is called a **parallel pencil**.

More generally, let $A\mathbf{x} = \mathbf{a}$ be the implicit and $\mathbf{x} = \mathbf{c} + C\mathbf{y}$ the parameter representation of some affine subspace $\mathcal{A}^n$. Again, if $\mathbf{a}$ and therefore also $\mathbf{c}$ vary, one obtains a family of affine subspaces with the same direction. This family has $n - r$ parameters and is called a **parallel bundle**. Figure 9.3 shows two fundamental examples in a 3-dimensional affine space.

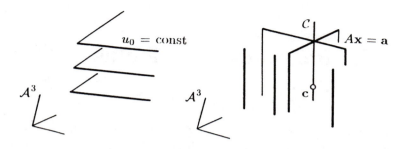

Figure 9.3: Parallel pencil and parallel bundle.

9.6 Notes and Problems

1 The empty affine space $\mathcal{A} = \emptyset$ has no underlying linear space. One defines $dim\ \mathcal{A} = -1$.

2 A single point forms an affine space of dimension 0. Its underlying linear space is the null space consisting only of **o**.

3 The affine intersection of two affine subspaces $\mathcal{A}$ and $\mathcal{B}$ equals the set-theoretical intersection $\mathcal{A} \cap \mathcal{B}$.

4 The direction of $\mathcal{A} \sqcap \mathcal{B}$ equals the intersection $\mathbf{A} \cap \mathbf{B}$ of the underlying linear spaces.

5 Two affine subspaces $\mathcal{A}$ and $\mathcal{B}$ are called **parallel**, denoted by $\mathcal{A} \parallel \mathcal{B}$, if $\mathbf{A} \subset \mathbf{B}$ or $\mathbf{B} \subset \mathbf{A}$.

6 $\mathcal{A}$ and $\mathcal{B}$ are called **skew** if the intersection of $\mathcal{A}$ and $\mathcal{B}$ is empty and $\mathcal{A}$ and $\mathcal{B}$ are not parallel.

10 The Barycentric Calculus

The underlying linear space $\mathbf{A}$ is used to define affine coordinates in $\mathcal{A}$. However, vectors are not necessary to describe points. In 1827 Möbius introduced barycentric coordinates which define a point with respect to some basis points. Barycentric coordinates are symmetric relative to these basis points, and they provide excellent insight into affine spaces and their structure.

Literature: Baker, Blaschke, Möbius

10.1 Barycentric Coordinates

Affine coordinates in an affine space $\mathcal{A}$ refer to a basis $\mathbf{a}_1, \ldots, \mathbf{a}_n$ of the underlying linear space $\mathbf{A}$. The vector basis can be suppressed by introducing the points $\mathbf{p}_0 = \mathbf{a}$, $\mathbf{p}_1 = \mathbf{a} + \mathbf{a}_1$, $\ldots$, $\mathbf{p}_n = \mathbf{a} + \mathbf{a}_n$. Thus, an arbitrary point $\mathbf{p} = \mathbf{a} + \mathbf{a}_1 x_1 + \cdots + \mathbf{a}_n x_n$ has the representation

$$\mathbf{p} = \mathbf{p}_0\left(1 - x_1 - \cdots - x_n\right) + \mathbf{p}_1 x_1 + \cdots + \mathbf{p}_n x_n$$

or more symmetrically

$$\mathbf{p} = \mathbf{p}_0 x_0 + \mathbf{p}_1 x_1 + \cdots + \mathbf{p}_n x_n \ ,$$

where x_0 is defined by

$$1 = x_0 + x_1 + \cdots + x_n \ .$$

Combining the last two equations one obtains

$$\begin{bmatrix} \mathbf{p} \\ 1 \end{bmatrix} = \begin{bmatrix} \mathbf{p}_1 & \cdots & \mathbf{p}_n & \mathbf{p}_0 \\ 1 & \cdots & 1 & 1 \end{bmatrix} \begin{bmatrix} \mathbf{x} \\ x_0 \end{bmatrix} .$$

This representation of $\mathbf{p}$ where the coefficients of the points sum to 1 is called an **affine combination**. The $x_0, \ldots, x_n$ are called **barycentric coordinates** of $\mathbf{p}$ with respect to the **frame** $\mathbf{p}_0, \ldots, \mathbf{p}_n$. One says that $\mathcal{A}$ is spanned by $\mathbf{p}_0, \ldots, \mathbf{p}_n$, or $\mathcal{A}$ is the **affine hull** of $\mathbf{p}_0, \ldots, \mathbf{p}_n$, more concisely $\mathcal{A} = \text{aff} [\mathbf{p}_0 \cdots \mathbf{p}_n]$. Note that the barycentric coordinates are unique.

Vectors can also be represented by barycentric coordinates. Let $\mathbf{q}$ be another point and $y_0, \ldots, y_n$ its barycentric coordinates. Then $x_i - y_i, i = 0, \ldots, n$, establish the barycentric coordinates of the vector $\mathbf{p} - \mathbf{q}$. Therefore a column of barycentric coordinates represents a vector if the coordinates sum to 0 and a point if they sum to 1.

Remark 1: Substituting $\mathbf{p}_i - \mathbf{p}_0$ for $\mathbf{a}_i$ only for $i = 0, \ldots, r$ yields

$$\mathbf{p} = \mathbf{p}_0 x_0 + \cdots + \mathbf{p}_r x_r + \mathbf{a}_{r+1} x_{r+1} + \cdots + \mathbf{a}_n x_n , \quad \text{where}$$
$$1 = \quad x_0 + \cdots + \quad x_r .$$

The vectors $\mathbf{a}_{r+1}, \ldots, \mathbf{a}_n$ can be viewed as points at infinity. In particular, $x_0, \ldots, x_n$ are affine coordinates for $r = 0$ and they are proper barycentric coordinates for $r = n$. This is illustrated in Figure 10.1.

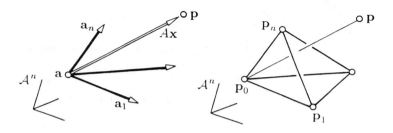

Figure 10.1: Affine system and barycentric system.

Remark 2: Barycentric coordinates are obtained from affine coordinates by adding the coordinate $x_0 = 1 - x_1 - \cdots - x_n$. Conversely, any n coordinates among $x_0, \ldots, x_n$ are affine coordinates.

10.2 Subspaces

Barycentric coordinates can also be introduced in an **affine subspace** $\mathcal{B} \subset \mathcal{A}$. Let $\mathcal{B}$ be spanned by $r+1$ basis points $\mathbf{q}_0, \ldots, \mathbf{q}_r$. Then every $\mathbf{x} \in \mathcal{B}$ has a unique representation

$$\mathbf{x} = \mathbf{q}_0 y_0 + \cdots + \mathbf{q}_r y_r \ ,$$
$$1 = \quad y_0 + \cdots + \quad y_r \ ,$$

where the $\mathbf{q}$'s and consequently $\mathbf{x}$ may be given with respect to some coordinate system of $\mathcal{A}$.

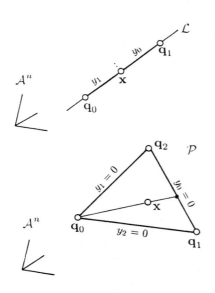

Example 1: An **affine line** $\mathcal{L}$ is represented by

$$\mathbf{x} = \mathbf{q}_0 y_0 + \mathbf{q}_1 y_1 \ ,$$
$$1 = \quad y_0 + \quad y_1 \ .$$

Rewriting $\mathbf{x}$ as

$$\mathbf{x} = \mathbf{q}_0 + y_1(\mathbf{q}_1 - \mathbf{q}_0)$$

shows that $\mathbf{x}$ divides the points $\mathbf{q}_0$ and $\mathbf{q}_1$ in the ratio $y_1 : y_0$.

Example 2: An **affine plane** $\mathcal{P}$ is represented by

$$\mathbf{x} = \mathbf{q}_0 y_0 + \mathbf{q}_1 y_1 + \mathbf{q}_2 y_2 \ ,$$
$$1 = \quad y_0 + \quad y_1 + \quad y_2 \ .$$

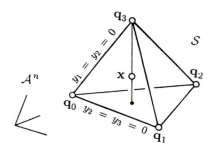

Example 3: An **affine space** S of dimension 3 is represented by

$$x = q_0 y_0 + q_1 y_1 + q_2 y_2 + q_3 y_3 \ ,$$
$$1 = \quad y_0 + \quad y_1 + \quad y_2 + \quad y_3 \ .$$

Example 4: Let $r = n$. Then the subspace $B \subset A$ coincides with A, and the $y_0, \ldots, y_n$ are the **barycentric coordinates** with respect to the **new frame** $q_0, \ldots, q_n$.

Remark 3: Let $p, p_0, \ldots, p_r$ be some points in B and v some vector of the underlying vector space. Then, the affine, extended or barycentric coordinates of the point $p + v$ are obtained by adding the affine, extended or respectively barycentric coordinates of p and v. Similarly, the affine combination $p = p_0 x_0 + \cdots + p_r x_r$, where $x_0 + \cdots + x_r = 1$, holds true (with the same weights $x_0, \ldots, x_r$) whether $p, p_0, \ldots, p_r$ are represented by affine, extended or barycentric coordinates.

10.3 Affine Independence

A family of $r + 1$ points of A, $q_0, \ldots, q_r$, is called **affinely independent** if the r vectors $b_1 = q_1 - q_0, \ldots, b_r = q_r - q_0$ are linearly independent. One can easily check that this definition does not depend on the choice of q_0.

The points $q_0, \ldots, q_r$ are affinely independent if their extended or barycentric coordinate columns are linearly independent and vice versa. Evidently, any $r + 1$ affinely independent points of the A^n can be extended to a frame of A^n. The converse procedure is of particular interest in some applications: If $x_r \neq 1$, the affine combination

$$p = p_0 x_0 + \cdots + p_r x_r \ ,$$
$$1 = \quad x_0 + \cdots + \quad x_r$$

can be written as

$$\mathbf{p} = \frac{\mathbf{p}_0 x_0 + \cdots + \mathbf{p}_{r-1} x_{r-1}}{x_0 + \cdots + x_{r-1}} \cdot \left(1 - x_r\right) + \mathbf{p}_r x_r$$

$$= \mathbf{q}_r \left(1 - x_r\right) + \mathbf{p}_r x_r \; ,$$

where

$$\mathbf{q}_r = \mathbf{p}_0 z_0 + \cdots + \mathbf{p}_{r-1} z_{r-1} \; , \quad z_i = \frac{x_i}{1 - x_r} \; .$$

Since $z_0, \ldots, z_{r-1}$ sum to 1, they are the barycentric coordinates of $\mathbf{q}_r$, the projection of $\mathbf{x}$ from $\mathbf{p}_r$ onto the plane spanned by $\mathbf{p}_0, \ldots, \mathbf{p}_{r-1}$. This is illustrated in Figure 10.2.

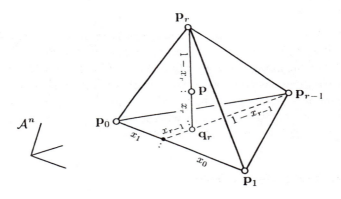

Figure 10.2: Reducing barycentric coordinates.

Example 5: Applying this reduction to a point

$$\mathbf{p} = \mathbf{a}\alpha + \mathbf{b}\beta + \mathbf{c}\gamma \; ,$$
$$1 = \alpha + \beta + \gamma$$

of a plane in different ways one obtains several ratios, which are depicted in Figure 10.3. The triangle in Figure 10.3 can be regarded as a part of the tetrahedron in Figure 10.2.

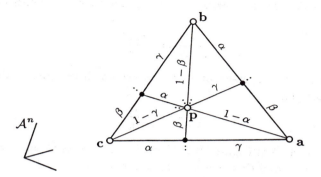

Figure 10.3: Ratios in a triangle.

10.4 Hyperplanes

A linear equation in affine coordinates of $\mathbf{x}$,

$$u(\mathbf{x}) = u_1 x_1 + \cdots + u_n x_n + u_0 = 0 \ ,$$

defines a hyperplane. Switching to barycentric coordinates one gets the additional coordinate x_0, defined by

$$b(\mathbf{x}) = x_0 + x_1 + \cdots + x_n - 1 = 0 \ .$$

Hence, the hyperplane can be represented with respect to the barycentric coordinates of $\mathbf{x}$ simply by the equation

$$u(\mathbf{x}) + \lambda b(\mathbf{x}) = 0 \ , \lambda \in \mathbb{R} \ .$$

For $\lambda = u_0$ the equation reads

$$v(\mathbf{x}) = v_0 x_0 + \cdots + v_n x_n = 0 \ ,$$

where $v_0 = u_0, v_1 = u_1 + u_0, \ldots, v_n = v_n + u_0$.

This equation is homogeneous in the barycentric coordinates of $\mathbf{x}$ and is called the **symmetric representation** of the hyperplane $\mathcal{U}$. A straight line

$\mathbf{x} = \mathbf{p}(1 - \alpha) + \mathbf{q}\alpha$ intersects $\mathcal{U}$ at the point $\mathbf{x}$, which corresponds to α satisfying

$$\frac{\alpha}{1 - \alpha} = -\frac{v(\mathbf{p})}{v(\mathbf{q})} \ .$$

In particular, for the intercept of the hyperplane $\mathcal{U}$ on the axis through $\mathbf{p}_i$ and $\mathbf{p}_j$, one gets

$$\frac{\alpha}{1 - \alpha} = -\frac{v_i}{v_j} \ .$$

This is illustrated in Figure 10.4. Note that the quotient may be negative in certain cases and that one or both of the v_j may vanish.

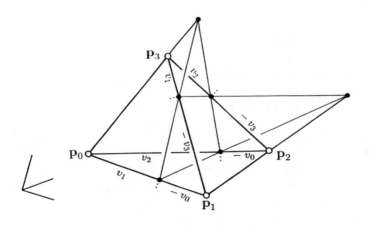

Figure 10.4: Intersections of a plane with the fundamental edges.

Example 6: In barycentric coordinates the plane $x_3 - 1 = 0$ in $\mathcal{A}^3$ has the symmetric representation $x_0 + x_1 + x_2 = 0$.

10.5 Join

Let $\mathcal{A} = \text{aff} \, [\mathbf{p}_0, \dots, \mathbf{p}_r]$ and $\mathcal{B} = \text{aff} \, [\mathbf{p}_{r+1} \cdots \mathbf{p}_s]$ be two affine subspaces of some larger space. In general, their set-theoretical union $\mathcal{A} \cup \mathcal{B}$ is not an affine space. Therefore the **join** $\mathcal{A} \sqcup \mathcal{B}$ is defined as the affine hull of the points $\mathbf{p}_0, \ldots, \mathbf{p}_s$, i.e., $\mathcal{A} \sqcup \mathcal{B} = \text{aff} \, [\mathbf{p}_0 \, \cdots \, \mathbf{p}_s]$. From the definition of the affine hull in Section 10.1, it follows that the join is an affine subspace.

The points $\mathbf{p}_0, \ldots, \mathbf{p}_s$ are not necessarily affinely independent. However, one can compute a frame of $\mathcal{A} \sqcup \mathcal{B}$ by an affine variant of the Gaussian elimination method applied to the rows $\mathbb{p}_i^t$, where the $\mathbb{p}_i$ represent either the extended or barycentric coordinate columns of the $\mathbf{p}_i$. After using Gaussian elimination to reduce the rows $\mathbb{p}_i^t$ to a linearly independent set of rows, one must multiply the rows by suitable non-zero factors to obtain extended or barycentric coordinates again. At this stage, some of the rows may represent vectors. However, there always is at least one row which represents a point. One can add this row to the rows representing vectors to obtain a frame for $\mathcal{A} \sqcup \mathcal{B}$. This procedure is illustrated below, with extended affine coordinates and $dim \, \mathcal{A} \sqcup \mathcal{B} = r$.

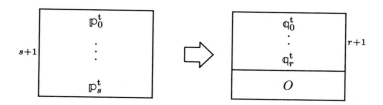

Note that the number of affinely independent points equals $dim \, \mathcal{A} \sqcup \mathcal{B} + 1$.

10.6 Volumes

The **convex hull** of a subset B of some affine space $\mathcal{A}$ is the smallest subset of $\mathcal{A}$ which contains B and which also contains the line segment connecting any two points of B. If B consists of $r + 1$ affinely independent points $\mathbf{q}_0, \ldots, \mathbf{q}_r$, the convex hull of B is called a **simplex**. For example, a simplex is a triangle if $r = 2$ and a tetrahedron if $r = 3$.

Let $\triangle$ be the volume of the simplex spanned by $\mathbf{q}_0, \ldots, \mathbf{q}_r$ and $\triangle_i$ the volume of this simplex, where $\mathbf{q}_i$ is replaced by

$$\mathbf{q} = \mathbf{q}_0 y_0 + \cdots + \mathbf{q}_r y_r, \quad 1 = y_0 + \cdots + y_r \, ,$$

as illustrated in Figure 10.5. One has $\sum \Delta_i = \Delta$ and, as can be seen from Section 10.2 and Figure 10.5,

$$y_r = \frac{\Delta_r}{\Delta} \; .$$

Hence, it is possible to compute the ratio of volumes in an affine space. Notice that the two volumes must be of the same dimension and must lie in parallel subspaces.

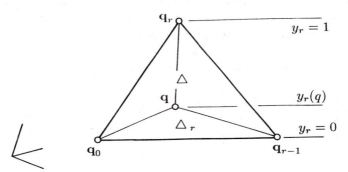

Figure 10.5: Ratios of volumes.

Remark 4: The frame $\mathbf{q}_0, \ldots, \mathbf{q}_r$ spans an affine subspace $\mathcal{B}$. Therefore one may assume that the $\mathbf{q}_i$ are given as r columns with respect to some affine coordinate system of $\mathcal{B}$. Hence, one can apply Cramer's rule to solve the linear system

$$\begin{bmatrix} \mathbf{q}_0 \cdots \mathbf{q}_r \\ 1 \ldots 1 \end{bmatrix} \mathbf{y} = \begin{bmatrix} \mathbf{q} \\ 1 \end{bmatrix}$$

for $\mathbf{y} = [y_0 \cdots y_r]^t$. As a result one obtains a formula for the volumes

$$\Delta_i = \varrho \det \begin{bmatrix} \mathbf{q}_0 \cdots \mathbf{q}_r \\ 1 \ldots 1 \end{bmatrix} \quad \text{and} \quad \Delta = \varrho \det \begin{bmatrix} \mathbf{q}_0 \cdots \mathbf{q}_r \\ 1 \ldots 1 \end{bmatrix},$$

where in the first determinant $\mathbf{q}_i$ is replace by $\mathbf{q}$, and ϱ is some scaling factor.

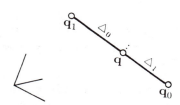

Example 7: If $r = 1$, the volumes $\triangle_0, \triangle_1$, and $\triangle$ are distances. In particular,

$$\mathbf{q} = (\mathbf{q}_0\triangle_0 + \mathbf{q}_1\triangle_1)/\triangle \, ,$$

as illustrated to the left.

10.7 A Generalization of Barycentric Coordinates

For certain applications it is necessary to define barycentric coordinates with respect to more than three points in the plane. This can be accomplished in the following way. Consider an n-gon with vertices $\mathbf{p}_1, \ldots, \mathbf{p}_n$ and an arbitrary point $\mathbf{p}$ in the plane. As illustrated in Figure 10.6, let $\triangle_{12}$ denote the area of the triangle $\mathbf{p}\,\mathbf{p}_1\mathbf{p}_2$, let $\triangle_2$ denote the area of the triangle $\mathbf{p}_1\mathbf{p}_2\mathbf{p}_3$, etc. Then one has

$$\mathbf{p}\sigma = \mathbf{p}_1\varphi_1 + \cdots + \mathbf{p}_n\varphi_n \, ,$$

where on taking indices $\mathrm{mod}\, n$

$$\varphi_i = \triangle_i\triangle_{(i+1)(i+2)} \cdots \triangle_{(i+n-2)(i+n-1)}$$

$$\text{and } \sigma = \varphi_1 + \cdots + \varphi_n \, .$$

Note that $\varphi_i/\triangle_i$ does not depend on $\mathbf{p}_i$ and that φ_i vanishes if $\mathbf{p}$ lies on a side opposite to $\mathbf{p}_i$. Obviously, this definition generalizes barycentric coordinates . In order to verify the representation of $\mathbf{p}$ above, express $\mathbf{p}$ with respect to $\mathbf{p}_{i-1}, \mathbf{p}_i$ and $\mathbf{p}_{i+1}$,

$$\mathbf{p}\triangle_i = \mathbf{p}_{i-1}\triangle_{i(i+1)} + \mathbf{p}_i(\triangle_i - \triangle_{(i-1)i} - \triangle_{i(i+1)}) + \mathbf{p}_{i+1}\triangle_{(i-1)i} \, .$$

Multiplying this by $\varphi_i/\triangle_i$ and summing the terms for all i, one gets the above representation for $\mathbf{p}$.

Consider n points $\mathbf{v}_1, \ldots, \mathbf{v}_n$ in the plane or in space and the surface defined by

$$\mathbf{q}(\mathbf{p}) = (\mathbf{v}_1\varphi_1 + \cdots + \mathbf{v}_n\varphi_n)/\sigma \ .$$

One has $\mathbf{q} = \mathbf{v}_1(1-\alpha) + \mathbf{v}_2\alpha$ if $\mathbf{p} = \mathbf{p}_1(1-\alpha) + \mathbf{p}_2\alpha$, etc., which implies that means that the surface meets the spatial n-gon spanned by the $\mathbf{v}_i$. Moreover, $\mathbf{p}_1 \sqcup \mathbf{p}_2$ is affinely related to $\mathbf{v}_1 \sqcup \mathbf{v}_2$, etc.

Since the $\triangle_{k(k+1)}$ are linear in $\mathbf{p}$, the φ_i are of degree $n-2$ while $\mathbf{q}$ is a rational polynomial of degree $n-2$. Moreover, φ_i cannot be of any lower degree, since the algebraic curve given by $\varphi_i(\mathbf{p}) = 0$ consists of $n-2$ lines.

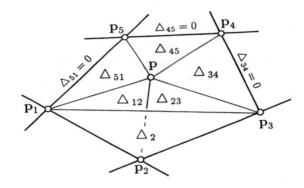

Figure 10.6: Barycentric coordinates in a quintangle.

10.8 Notes and Problems

1 The set theoretical union $\mathcal{A} \cup \mathcal{B}$ is a proper subset of the join $\mathcal{A} \sqcup \mathcal{B}$, unless $\mathcal{B}$ is a subspace of $\mathcal{A}$ or $\mathcal{A}$ a subspace of $\mathcal{B}$.

2 Let $\mathbf{A}$ and $\mathbf{B}$ be the linear spaces associated with two affine subspaces $\mathcal{A}$ and $\mathcal{B}$. Then the dimensions of $\mathcal{A}$, $\mathcal{B}$, of the intersection and the join of $\mathcal{A}$ and $\mathcal{B}$ are related by the **dimension theorem**

$$dim\, \mathcal{A} + dim\, \mathcal{B} = dim\, \mathcal{A} \sqcap \mathcal{B} + \begin{cases} dim\, \mathcal{A} \sqcap \mathcal{B} & \text{if } \mathcal{A} \sqcap \mathcal{B} \neq \emptyset \\ dim\, \mathbf{A} \sqcap \mathbf{B} - 1 & \text{if } \mathcal{A} \sqcup \mathcal{B} = \emptyset \end{cases} .$$

3 The **dimension theorem for linear spaces** states that

$$dim\, \mathbf{A} + dim\, \mathbf{B} = dim\, \mathbf{A} \sqcup \mathbf{B} + dim\, \mathbf{A} \sqcap \mathbf{B} .$$

4 The intersection of all affine subspaces of some affine space $\mathcal{A}$ containing a point set $Q \subset \mathcal{A}$ is the affine hull of Q.

5 The join of two affine subspaces is their affine hull.

6 Let α, β, γ be the barycentric coordinates of a point in some plane with respect to the frame $\mathbf{a}, \mathbf{b}, \mathbf{c}$. Figure 10.7 shows the isolines where α, β or γ is equal to an integer constant. The isolines form a lattice which consists of translates of the basis triangle $\triangle$ and triangular gaps which also are translates of each other.

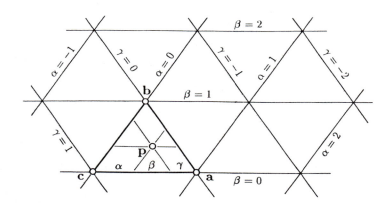

Figure 10.7: Barycentric isolines in the plane

7 Figure 10.8 shows the isoplanes where one of the barycentric coordinates in $\mathcal{A}^3$ is equal to an integer constant. The isoplanes form a lattice which consists of translates of the basis tetrahedron and octahedral gaps which also are translates of each other.

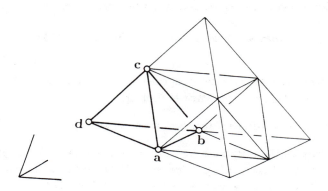

Figure 10.8: Barycentric isoplanes in the space.

8 Let q_i denote the additional barycentric coordinate of a point with affine coordinates $\mathbf{q}_i$ in A^r. Then

$$\det \begin{bmatrix} \mathbf{q}_0 \cdots \mathbf{q}_r \\ 1 \ldots 1 \end{bmatrix} = \det \begin{bmatrix} \mathbf{q}_0 \cdots \mathbf{q}_r \\ q_0 \cdots q_r \end{bmatrix} .$$

11 Affine Maps

Maps mapping affine spaces into affine spaces are of particular interest if they preserve the affine structure, i.e., if they preserve affine combinations. Such maps are called affine and are also characterized by the property that they induce a linear map between the underlying vector spaces. Like linear maps, affine transformations have a matrix representation. Examples of affine maps are the parallel projections in Sections 4.1–4.3.

Literature: Baker, Berger, Schaal

11.1 Barycentric Representation

Consider two affine spaces $\mathcal{A}$ and $\mathcal{B}$ of dimensions n and m, respectively, and a map $\Phi : \mathcal{A} \to \mathcal{B}$ which leaves affine combinations invariant. The transformation Φ is called an **affine map**. Let $\mathbf{p}_0, \ldots, \mathbf{p}_n$ form a barycentric coordinate system of $\mathcal{A}$ and let $\mathbf{q}_0, \ldots, \mathbf{q}_n$ be its image in $\mathcal{B}$, as illustrated in Figure 11.1. Since Φ is affine, a point $\mathbf{x} = \mathbf{p}_0 x_0 + \cdots + \mathbf{p}_n x_n$ of $\mathcal{A}$ is mapped onto the point

$$\mathbf{y} = \mathbf{q}_0 x_0 + \cdots + \mathbf{q}_n x_n$$

of $\mathcal{B}$, i.e., one has

$$\begin{bmatrix} \mathbf{y} \\ 1 \end{bmatrix} = \begin{bmatrix} \mathbf{q}_1 & \cdots & \mathbf{q}_n & \mathbf{q}_0 \\ 1 & \cdots & 1 & 1 \end{bmatrix} \begin{bmatrix} \mathbf{x} \\ x_0 \end{bmatrix} .$$

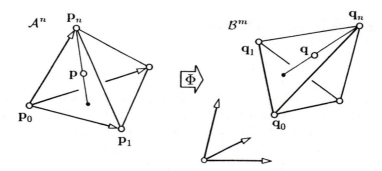

Figure 11.1: Definition of an affine map.

An immediate consequence of this equation is the following **theorem**:

> An affine map $\Phi : \mathcal{A} \to \mathcal{B}$ is uniquely determined by the images of $dim\, \mathcal{A} + 1$ affinely independent points of $\mathcal{A}$.

Note that the images $\mathbf{q}_0, \dots, \mathbf{q}_n$ neither have to be affinely independent nor do they have to span $\mathcal{B}$.

In the discussion above the points $\mathbf{q}_0, \dots, \mathbf{q}_n$ and $\mathbf{y}$ are represented by their affine coordinates. However, they also could have been represented by their barycentric coordinates. (This is also observed in Section 10.2, Remark 3.) Therefore if $q_0, \dots, q_n$ and y denote the additional barycentric coordinates of $\mathbf{q}_0, \dots, \mathbf{q}_n$ and $\mathbf{y}$, respectively, Φ can be written as

$$\begin{bmatrix} \mathbf{y} \\ y \end{bmatrix} = \begin{bmatrix} \mathbf{q}_1 \cdots \mathbf{q}_n \mathbf{q}_0 \\ q_1 \cdots q_n\, q_0 \end{bmatrix} \begin{bmatrix} \mathbf{x} \\ x_0 \end{bmatrix},$$

or more concisely as $y = Q\mathbf{x}$.

Remark 1: In the case where $m = n$ and the $\mathbf{q}_i$ are affinely independent, $\mathbf{q}_0, \dots, \mathbf{q}_n$ forms a frame of $\mathcal{B}$. If this frame has been chosen as the coordinate system of $\mathcal{B}$, the matrix Q is the identity and $y = \mathbf{x}$, i.e., the points of $\mathcal{A}$ and $\mathcal{B}$ are said to be **related by common coordinates**.

Remark 2: If $\mathbf{x}$ and $\mathbf{p}_0, \ldots, \mathbf{p}_n$ are given with respect to some coordinate system different from $\mathbf{p}_0, \ldots, \mathbf{p}_n$, the map Φ is given by $\mathsf{y} = \mathbb{Q}\mathbb{P}^{-1}\mathsf{x}$ where

$$\mathbb{P} = \begin{bmatrix} \mathbf{p}_1 \cdots \mathbf{p}_n \mathbf{p}_0 \\ p_1 \cdots p_n \ p_0 \end{bmatrix} .$$

11.2 Affine Representation

On returning to affine coordinates by substituting $1 - x_1 - \cdots - x_n$ for x_0, one obtains from the previous section

$$\mathbf{y} = \mathbf{q}_0 + (\mathbf{q}_1 - \mathbf{q}_0)x_1 + \cdots + (\mathbf{q}_n - \mathbf{q}_0)x_n \ ,$$

where $x_1, \ldots, x_n$ are the affine coordinates of $\mathbf{p}$ with respect to the affine system of $\mathcal{A}$ given in Section 10.1.

With the abbreviations $\mathbf{c} = \mathbf{q}_0$, $\mathbf{c}_i = \mathbf{q}_i - \mathbf{q}_0$, and $C = [\mathbf{c}_1 \ldots \mathbf{c}_n]$ one obtains the affine representation of Φ

$$\mathbf{y} = \mathbf{c} + C\mathbf{x} \ .$$

The differences $\mathbf{c}_i$ are the images of the vectors $\mathbf{a}_i$ under Φ. Moreover, since every vector $\triangle\mathbf{x}$ is the difference $\mathbf{x}_2 - \mathbf{x}_1$ of two points, its image $\triangle\mathbf{y}$ under Φ is the difference $\Phi\mathbf{x}_2 - \Phi\mathbf{x}_1$ of the image points, i.e., one has

$$\triangle\mathbf{y} = C\triangle\mathbf{x}$$

which represents the linear map associated with Φ.

Note that the images $\mathbf{c}_1, \ldots, \mathbf{c}_n$ neither have to be linearly independent nor do they have to span the underlying linear space of $\mathcal{B}$. One can observe that $\Phi\mathcal{A}$ is an affine space with

$$dim\,(\Phi\mathcal{A}) = rank\,C \ .$$

11.3 Parallelism and Ratio

Let $\mathbf{x}_1, \mathbf{x}_2$ and $\mathbf{z}_1, \mathbf{z}_2$ be the coordinates of two pairs of points in $\mathcal{A}$. These pairs lie on parallel lines and possess the **ratio** ϱ if

$$\triangle \mathbf{x} = \varrho \triangle \mathbf{z} \neq \mathbf{o} \ ,$$

where $\triangle \mathbf{x} = \mathbf{x}_2 - \mathbf{x}_1$ and $\triangle \mathbf{z} = \mathbf{z}_2 - \mathbf{z}_1$. Consequently the images of both pairs in $\mathcal{B}$ also lie on parallel lines and have the same ratio ϱ, if $C\triangle \mathbf{x} \neq \mathbf{o}$. Thus, one has the following theorem:

Affine maps preserve parallelism and ratios.

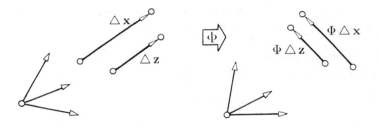

Figure 11.2: Parallelism and ratio.

Remark 3: The converse is also true: Any map which preserves parallelism and ratios is an affine map. This fact has already been used in Chapter 4.

11.4 Fibers

The $n + 1$ image points $\mathbf{q}_i$ span the image $\Phi \mathcal{A}$ of $\mathcal{A}$ which is an affine subspace of $\mathcal{B}$ with $dim\, \Phi \mathcal{A} \leq dim\, \mathcal{A}$. The possible loss of dimensions can be analyzed as follows. The preimage of some point $\mathbf{q} \in \Phi \mathcal{A}$ is given by the solution $\mathcal{F}$ of the linear system

$$\mathbf{q} = \mathbf{c} + C\mathbf{x} \ .$$

This means that $\mathcal{F}$ is an affine subspace whose direction $\mathbf{F}$ is represented by the solution of the underlying homogeneous system

$$C \triangle \mathbf{x} = \mathbf{o}$$

which does not depend on $\mathbf{q}$. Because of Section 2.4 one has

$$dim\ \Phi\mathcal{A} + dim\ \mathcal{F} = dim\ \mathcal{A}\ .$$

The affine subspace $\mathcal{F}$ is called a **fiber** of Φ. All fibers of Φ are parallel and have the same dimension. They cover $\mathcal{A}$ exactly once.

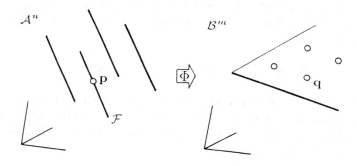

Figure 11.3: Fibers of an affine map.

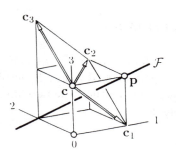

Example 1: Let $\Phi : \mathcal{A}^3 \rightarrow \mathcal{A}^3$ be given by

$$\mathbf{y} = \begin{bmatrix} 0 \\ 0 \\ 1 \end{bmatrix} + \begin{bmatrix} 1 & 1 & 0 \\ 0 & 1 & 1 \\ -1 & 0 & 1 \end{bmatrix} \mathbf{x}\ .$$

The fiber $\mathcal{F}$ of Φ containing $\mathbf{p}$ is

$$\mathbf{x} = \mathbf{p} + \begin{bmatrix} 1 \\ -1 \\ 1 \end{bmatrix} \lambda\ .$$

In the example above, $dim\,\mathcal{F} = 1$ and $dim\,\Phi\mathcal{A} = 2$. The family of fibers consists of all affine lines in the direction $\mathbf{v} = [1\ \ -1\ \ 1]^t$.

11.5 Affinities

An affine map Φ of an affine space $\mathcal{A}$ into itself is called an **affinity**; if the map is onto it is called a **regular affinity**. The matrix C of an affinity Φ is square and, if and only if Φ is regular, also non-singular.

Of particular interest are fixed points or fixed directions of affinities. Provided that images and preimages are given with respect to the same coordinate systems, fixed points and fixed directions are the respective solutions of the linear systems

$$\mathbf{x} = \mathbf{c} + C\mathbf{x} \qquad \text{and} \qquad C\triangle\mathbf{x} = \triangle\mathbf{x}\,\varrho \ .$$

A map $\Psi : \mathcal{A} \to \mathcal{A}$ is **idempotent** if $\Psi \circ \Psi = \Psi$. Idempotent affine maps are called **projections**. Projections are also characterized by the property that their fibers contain their images. The parallel projections of Figure 4.5 may serve as examples.

Example 2: The affinity of Example 1 is non-regular. It has $[1\ \ \ 0\ \ \ 0]^t$ as a fixed point but no fixed directions.

Example 3: One has a translation for $C = I$ but $\mathbf{c} \neq \mathbf{o}$. A translation has no fixed points while all directions are fixed.

11.6 Correspondence of Hyperplanes

An affine map $\Phi : \mathcal{A} \to \mathcal{B}$ maps points into points and subspaces into subspaces, but, in general, it does not map hyperplanes of $\mathcal{A}$ into hyperplanes of $\mathcal{B}$. However, the preimage of a hyperplane is either a hyperplane, the entire space $\mathcal{A}$ or does not exist. Namely, let

$$\mathbf{v}^t\mathbf{y} + v_0 = 0$$

represent some hyperplane $\mathcal{V}$ of $\mathcal{B}$ and $\mathbf{y} = \mathbf{c} + C\mathbf{x}$ an affine map Φ. Then, the preimage of $\mathcal{V}$ is obtained as the solution of

$$\mathbf{v}^t(\mathbf{c} + C\mathbf{x}) + v_0 = \mathbf{u}^t\mathbf{x} + u_0 = 0 \ ,$$

where

$$\mathbf{u}^t = \mathbf{v}^t C \qquad \text{and} \qquad u_0 = \mathbf{v}^t\mathbf{c} + v_0 \ .$$

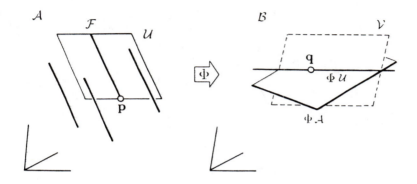

Figure 11.4: The correspondence of hyperplanes.

Moreover, Φ defines a correspondence between certain hyperplanes of $\mathcal{B}$ and $\mathcal{A}$ with the following geometric properties:

If $\mathcal{V} \sqcap \Phi\mathcal{A}$ is a proper subspace of $\Phi\mathcal{A}$, then $\mathbf{u} \neq \mathbf{o}$ and $\mathcal{U}$ is uniquely defined by $\mathcal{V}$. It is parallel to the fibers of Φ and mapped onto $\mathcal{V} \sqcap \Phi\mathcal{A}$.

If $\Phi\mathcal{A} \subset \mathcal{V}$, then $\mathbf{u} = \mathbf{o}$ and $u_0 = 0$, i.e., the preimage of $\mathcal{V}$ is the entire space $\mathcal{A}$.

If $\mathcal{V} \sqcap \Phi\mathcal{A}$ is empty, i.e., if $\mathcal{V}$ is parallel to $\Phi\mathcal{A}$, then $\mathbf{u} = \mathbf{o}$ but $u_0 \neq 0$. A preimage of $\mathcal{V}$ does not exist.

11.7 Notes and Problems

1 Occasionally it is convenient to define $\mathbf{o}$ to be parallel to all directions. Any two affine spaces of the same dimension can be viewed as affine images of each other.

2 Let $\Phi : \mathcal{A} \to \mathcal{B}$ and $\Psi : \mathcal{B} \to \mathcal{C}$ be affine maps. Then the composition $\Psi \circ \Phi : \mathcal{A} \to \mathcal{C}$ is also an affine map.

3 Let $\Phi : \mathcal{A} \to \mathcal{B}$ be an affine map and $\Psi : \mathcal{B} \to \mathcal{A}$ some arbitrary map. If $\Phi \circ \Psi$ is the identity on $\mathcal{B}$, then Ψ is called a **pseudo affine map**. An example of a pseudo affine map is shown in Figure 11.5.

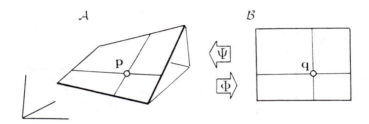

Figure 11.5: Pseudoaffine map.

4 Another example of a pseudo affine map can be found in Section 10.7 on the generalization of barycentric coordinates. If the $\mathbf{v}_i$'s span $\mathcal{A}$ and are independent, then $\Psi(\mathbf{p}) = \mathbf{q}(\mathbf{p})$ is pseudo affine. The map Ψ composed with the affine map which maps the $\mathbf{v}_i$ onto the corresponding $\mathbf{p}_i$ gives the identity.

5 Relative to a proper subspace $\mathcal{S}$ of $\mathcal{B}$ the hyperplanes of $\mathcal{B}$ fall into one of three categories: Hyperplanes intersecting $\mathcal{S}$ in a hyperplane of $\mathcal{S}$, hyperplanes containing $\mathcal{S}$, and hyperplanes disjoint from $\mathcal{S}$. Hyperplanes disjoint from $\mathcal{S}$ are parallel to $\mathcal{S}$.

6 The reconstruction of an object from some perspective images can be carried over to the construction of the object from some affine image if the mapping matrices are known.

7 The camera calibration from a perspective image of some known object can be carried over to the determination of the mapping matrix of a parallel projection including a scaling.

8 The determination of the mapping matrix of a parallel projection reflects Pohlke's theorem.

12 Affine Figures

Affine maps preserve the structure of the mapped affine spaces. Of particular interest are the properties which remain invariant when figures undergo affine mappings. These properties depend on the affine rules used to construct a figure not on the position of the figure in space. Often a special position allows for a simple proof of a general theorem. A pair of points and their midpoint form an example of a simple affine figure. The Bézier and B-spline representation of curves have affine properties which are rather intriguing and most crucial for geometric design.

Literature: Blaschke, Coxeter, Pedoe

12.1 Triangles

A number of classical and useful theorems can be verified by the relationships illustrated in Figure 10.3:

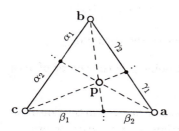

Ceva's Theorem

The dashed lines of the figure intersect at a common point p if and only if

$$\frac{\alpha_1}{\alpha_2} \cdot \frac{\beta_1}{\beta_2} \cdot \frac{\gamma_1}{\gamma_2} = 1 \ .$$

For a proof observe that each ratio $\alpha_1 : \alpha_2$ corresponds to a unique point

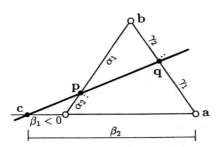

on **bc**. Hence, Figure 10.3 asserts that the product above equals 1 if and only if $\alpha_1 : \alpha_2$ corresponds to the black point on **bc**.

Menelaus' Theorem

The point **q** of the figure to the left is collinear with **c** and **p** if and only if

$$\frac{\alpha_1}{\alpha_2} \cdot \frac{\beta_1}{\beta_2} \cdot \frac{\gamma_1}{\gamma_2} = -1 \ .$$

This theorem can also be derived from Figure 10.3. Each ratio $\gamma_1 : \gamma_2$ corresponds to a unique point on **ab**. Hence, the product above equals -1 if and only if $\gamma_1 : \gamma_2$ corresponds to **q**. Note that in this case either one ratio or all three ratios must be negative.

A-frame Theorem

The ratios $\alpha_1 : \alpha_0$ and $\gamma_1 : \gamma_2$ of the figure to the left are given by

$$\alpha_1 : \alpha_0 = \Delta_0 : (\Delta_1 + \Delta_2) \quad \text{and}$$
$$\gamma_0 : \gamma_1 = (\Delta_0 + \Delta_1) : \Delta_2 \ .$$

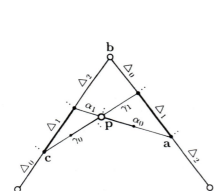

Using the notation of Figure 10.3 these two identities are

$$\alpha : (1 - \alpha) = \alpha : (\beta + \gamma) \quad \text{and}$$
$$(1 - \gamma) : \gamma = (\alpha + \beta) : \gamma,$$

which proves the theorem.

Remark 1: In case $\alpha = \beta = \gamma = 1/3$, Ceva's point **p** is the **center of area** of the triangle **a**, **b**, **c**.

Remark 2: An application of the A-frame theorem is given in Section 12.3. This theorem is important for many constructions in geometric design.

12.2 Quadrangles

In general, two planar quadrangles are not affine images of each other. Nonetheless, there are intriguing affine properties which are true for all planar quadrangles.

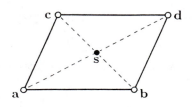

Parallelograms
Two pairs of opposite points **a**, **d** and **b**, **c** determine a **parallelogram** if and only if

$$\mathbf{a} + \mathbf{d} = \mathbf{b} + \mathbf{c} = 2\mathbf{s}\ .$$

The point **s** is the **center of area**.

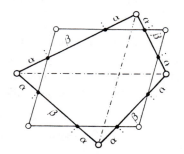

Corner Cutting
One can construct a parallelogram from any planar quadrangle by cutting all its edges in the same ratio $\alpha : \beta : \alpha$, as shown in the figure to the left.
Note that the quadrangle may even be non-convex.

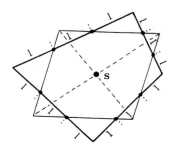

Wittenbauer's Theorem
Cutting all edges of a planar quadrangle in the ratio 1 : 1 : 1 yields a parallelogram with the same **center of area**.

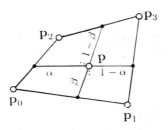

Biaffine Combinations

The **biaffine combination**

$$\mathbf{p} = [\mathbf{p}_0(1-\alpha) + \mathbf{p}_1\alpha](1-\beta)$$
$$+ [\mathbf{p}_2(1-\alpha) + \mathbf{p}_3\alpha]\beta$$

can also be written as

$$\mathbf{p} = [\mathbf{p}_0(1-\beta) + \mathbf{p}_2\beta](1-\alpha)$$
$$+ [\mathbf{p}_1(1-\beta) + \mathbf{p}_3\beta]\alpha .$$

Its geometric meaning is illustrated to the left.

Hyperbolic Paraboloid

Let $\mathbf{p}_0^* = \mathbf{p}_1 + \mathbf{p}_2 - \mathbf{p}_0$. The affine coordinates of the biaffine combination $\mathbf{p}$, with respect to the affine system $\mathbf{p}_0; \mathbf{p}_1 - \mathbf{p}_0, \mathbf{p}_2 - \mathbf{p}_0, \mathbf{p}_3 - \mathbf{p}_0^*$, are $x_1 = \alpha, x_2 = \beta, x_3 = \alpha\beta$.

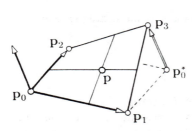

Note that these coordinates of $\mathbf{p}$ satisfy the equation $x_3 = x_1 x_2$. If $\mathbf{p}_0, \ldots, \mathbf{p}_3$ are affinely independent, this equation represents a **hyperbolic paraboloid**.

Remark 3: For $\mathbf{p}_1 = \mathbf{p}_2$ the property of the biaffine combination mentioned above implies the A-frame theorem.

12.3 Polygons and Curves

One of the very important properties of the so-called **Bernstein-Bézier representation** for polynomial curves is its affine invariance.

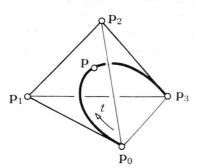

Bézier Curves

Since the **Bernstein polynomials**

$$B_i^n(t) = \binom{n}{i}(1-t)^{n-i}t^i$$

sum to 1, they can be used to form an affine combination of $n+1$ points $\mathbf{p}_0, \ldots, \mathbf{p}_n$,

$$\mathbf{p}(t) = \sum_{i=0}^{n} \mathbf{p}_i B_i^n(t) \ .$$

If t varies, $\mathbf{p}(t)$ traces out a polynomial curve of degree n. A polynomial curve with this representation is called a **Bézier curve**. The **Bézier points** $\mathbf{p}_i$ control the curve $\mathbf{p}(t)$.

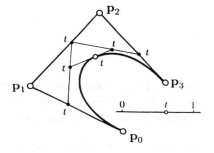

De Casteljau's Algorithm

Every point $\mathbf{p}(t)$ of a Bézier curve can be constructed by repeated affine combinations. This algorithm is due to **de Casteljau**. Let $\mathbf{p}_i^0 = \mathbf{p}_i$ and

$$\mathbf{p}_i^r = \mathbf{p}_i^{r-1}(1-t) + \mathbf{p}_{i+1}^{r-1}t$$

for $r = 1, \ldots, n$, then $\mathbf{p}_0^n = \mathbf{p}(t)$.

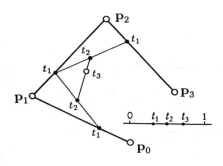

Blossom of a Bézier Curve

One may choose a different parameter t at each level r of de Casteljau's algorithm. Then one has

$$\mathbf{p}_i^r = \mathbf{p}_i^{r-1}(1-t_r) + \mathbf{p}_{i+1}^{r-1}t_r \ .$$

From the A-frame theorem one can infer that interchanging t_r with t_{r-1} does not change the points $\mathbf{p}_i^r$. Consequently, $\mathbf{p}_0^n$ does not depend on the ordering of the t_r. Some authors call $\mathbf{p}_0^n$ the **blossom** of $\mathbf{p}(t)$ at $t_1, \ldots, t_n$, see also Chapter 27.

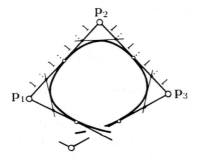

Chaikin's Algorithm

Repeated corner cutting of a polygon in the ratio 1 : 2 : 1 generates a smooth curve in the limit. This curve connects the midpoints of the polygon's edges by parabolic arcs so that the polygon's edges are tangential to the curve, as illustrated to the left.

12.4 Conic Sections

Conic sections, i.e., ellipses, hyperbolas, and parabolas, have a number of affine properties; for example, the affine image of an ellipse is an ellipse, that of a hyperbola is a hyperbola, and that of a parabola is a parabola.

Every **ellipse** can be viewed as an affine image of the **"unit circle"** $x^2 + y^2 = 1$. This view can be helpful when constructing points and tangents of an ellipse, e.g., as in the tangent octagon in Figure 12.1.

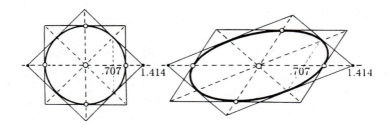

Figure 12.1: Tangent octagon of an ellipse.

Every **hyperbola** can be viewed as an affine image of the **"unit hyperbola"** $x^2 - y^2 = 1$ or of the simple hyperbola $y = 1/x$. Therefore the following two properties of these hyperbolas are shared by all hyperbolas: The areas bounded by the tangents and asymptotes of the hyperbola are all equal, and each tangent contacts the hyperbola at the midpoint of the segment connecting the two places in which the tangent and the asymptotes intersect.

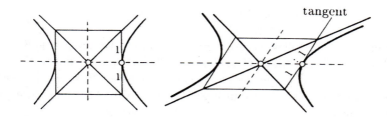

Figure 12.2: Affine image of the unit hyperbola.

Every **parabola** can be viewed as an affine image of the **"unit parabola"** $y = x^2$, where the y-direction is mapped onto the **axis direction**. Therefore the following property of the unit parabola is shared by all parabolas:

Let $\mathbf{p}_1, \mathbf{p}_2$ be two points of a parabola, $\mathbf{s}$ the intersection of the corresponding tangents, and $\mathbf{m}$ the midpoint of $\mathbf{p}_1$ and $\mathbf{p}_2$. Then the line $\mathbf{sm}$ is in the axis direction and meets the parabola at the midpoint of $\mathbf{s}$ and $\mathbf{m}$. The tangent at this intersection is parallel to the chord $\mathbf{p}_1$, $\mathbf{p}_2$. This configuration coincides with de Casteljau's construction for $n = 2$ and $t = 1/2$.

Similarly, one can use two special tangents of the unit parabola to verify the following important theorem:

The joins of affinely related points of two lines in general position envelope a parabola.

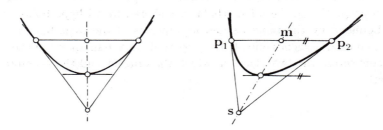

Figure 12.3: Affine image of the unit parabola.

Remark 4: The following property common to all conic sections can be verified in the same way. Consider a family of parallel chords of a conic section. The pairs of tangents at the chord ends meet at points on the line which bisects the chords, as illustrated in Figure 12.4 for an ellipse. Note that a tangent can be viewed as a chord of length 0.

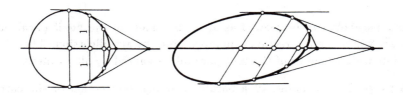

Figure 12.4: Parallel chords of an ellipse.

12.5 Axial Affinities

If an affinity $\mathcal{A}^n \rightarrow \mathcal{A}^n$ leaves a hyperplane $\mathcal{H}$ fixed pointwise, then the affinity is said to be **axial**. Let $\mathcal{H}$ be spanned by the points $\mathbf{p}_1, \ldots, \mathbf{p}_n$, and let $\mathbf{q}_0$ be the image of some $\mathbf{p}_0 \notin \mathcal{H}$. Then every point $\mathbf{p} \in \mathcal{A}^n$ has a representation

$$\mathbf{p} = \mathbf{p}_0 x_0 + \mathbf{p}_1 x_1 + \cdots + \mathbf{p}_n x_n$$

and is mapped into

$$\mathbf{q} = \mathbf{q}_0 x_0 + \mathbf{p}_1 x_1 + \cdots + \mathbf{p}_n x_n \; .$$

Consequently, one has

$$\mathbf{q} = \mathbf{p} + [\mathbf{q}_0 - \mathbf{p}_0] x_0 \; ,$$

where $\mathbf{v} = \mathbf{q}_0 - \mathbf{p}_0$ is called the **direction of the affinity**. $\mathcal{H}$ is called the **axis** or **axial plane** of the affinity. All points $\mathbf{p}$ with the same (fixed) x_0 form a hyperplane parallel to $\mathcal{H}$ which is translated by $\mathbf{v} x_0$, as illustrated in Figure 12.5.

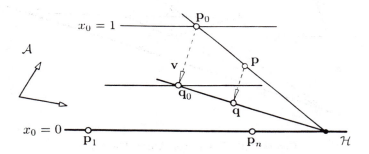

Figure 12.5: Axial affinity.

Example 1: Figure 12.6 shows two triangles in axial affine positions in the plane. This configuration exhibits **Desargues' Affine Theorem**:

> If the corresponding vertices of two triangles span three parallel lines, then the corresponding edges intersect in points of a straight line.

A proof is given in Section 22.2. Note that one can also allow ideal points.

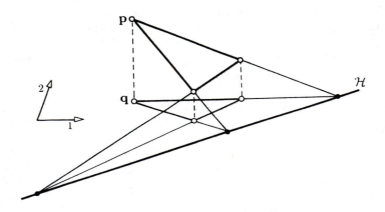

Figure 12.6: Desargues' Affine Theorem.

Example 2: Figure 12.7 shows two ellipses in two planes, which are in axial affine position. By selecting some points, one can demonstrate the affinity and vividly describe the relationship between the two shapes: Tangents correspond to tangents, midpoints to midpoints, etc. Note that the intersection of the two planes spanned by the ellipses lies in $\mathcal{H}$.

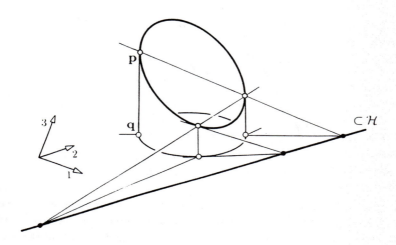

Figure 12.7: Axial affinity of two ellipses.

Remark 5: A pair of axial affine figures in the plane can always be viewed as a figure in space together with its parallel projection into another plane, as indicated in Figures 12.6 and 12.7.

12.6 Dilatation

An affinity which leaves all lines through some point c fixed is called a **dilatation** with center c. A dilatation stretches all directions by the same amount ϱ, i.e., a point $p = c + v$ is mapped onto $q = c + \varrho v$, as illustrated in Figure 12.8. Hence this dilatation is given by

$$q = c(1 - \varrho) + p\varrho ,$$

while the underlying linear map is given by

$$\triangle q = \triangle p\, \varrho .$$

Note that ϱ can be negative and that c can be determined by a pair p, q and the dilatation factor ϱ.

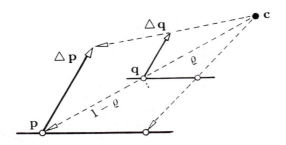

Figure 12.8: Dilatation.

Example 3: Two parallelograms constructed by corner cutting, as in Section 12.2, from the same quadrilateral but with different ratios are centric affine. The center agrees with the intersection of the diagonals of the given quadrangle. Let $2\alpha_1 + \beta_1 = 2\alpha_2 + \beta_2 = 1$ then $\varrho = (1 - \alpha_1) : (1 - \alpha_2)$.

12.7 Notes and Problems

1 Each diagonal of the quadrangle in the corner cutting construction of Section 12.2 is parallel to two edges of the resulting parallelogram. In proving Wittenbauer's theorem it is advantageous to use these diagonals as the axes of the affine system.

2 The fixed hyperplane of an axial affinity may be ideal. Then the affinity is a translation.

3 The center of a dilatation may be an ideal point. This affinity, then, is also a translation.

4 For $0 \leq t \leq 1$ one has $B_i^n(t) \geq 0$. As a consequence the respective segment of a Bézier curve lies in the **convex hull** of its control points.

5 Figure 12.8 can be regarded as an example of Desargues' General Theorem where the corresponding edges of both triangles meet in points of an ideal line.

6 Originally, Ceva and Menelaus used ratios with signs opposite to the ones above. Consequently the products of the ratios also had different signs.

7 An axial affinity is called a **shearing** if some point not in $\mathcal{H}$ and its image lie on a line which is parallel to $\mathcal{H}$.

8 In general, the center of area of a planar n-gon differs from the **center of gravity** of the n vertices, if $n > 3$.

9 A regular affinity leaves the ratio of two r-volumes in parallel subspaces of dimension r invariant.

13 Quadrics in Affine Spaces

The simplest figures in an affine space besides lines and planes are conics which are the intersection curves of planes and right circular cones. Conics were studied by the Greeks, mainly by Menaichmos (about 350 B.C.) and by Apollonios (200 B.C.), who introduced the names ellipse, hyperbola, and parabola. Conics can be conveniently studied using their quadratic equations, and, without additional effort, the analysis of these quadratic equations can be presented for general quadratic surfaces in any dimension, the so-called quadrics. In this chapter affine concepts of quadrics, such as midpoints, singular points, tangents, asymptotes, and polar planes are discussed.

Literature: Berger, Meserve, Samuel

13.1 The Equation of a Quadric

A **quadric** consists of all points $\mathbf{x}$ in an affine space $\mathcal{A}^n$ satisfying a quadratic equation which can be written as

$$Q(\mathbf{x}) = \mathbf{x}^{\mathrm{t}} C \mathbf{x} + 2\mathbf{c}^{\mathrm{t}}\mathbf{x} + c = 0 \ ,$$

where $C = C^{\mathrm{t}}$ is a symmetric non-zero $n \times n$ matrix. The quadric described by the equation $Q(\mathbf{x}) = 0$ will also be denoted by Q. The equation can be visualized by blocks:

$$\boxed{\begin{array}{c}\mathbf{x^t}\\ n\end{array}}\ \boxed{\begin{array}{c} C \\ n\end{array}}_{n}\boxed{\mathbf{x}} + \boxed{2}\ \boxed{\begin{array}{c}\mathbf{c^t}\\ n\end{array}}\ \boxed{\mathbf{x}} + \boxed{c} = \boxed{0}\ .$$

The intersection of Q with an affine subspace $\mathcal{B}$ of dimension $r \geq 1$ is a quadric again or a hyperplane in $\mathcal{B}$, i.e., a subspace of dimension $r - 1$. To prove this let $\mathcal{B}$ be represented by

$$\mathbf{x} = \mathbf{b} + B\mathbf{y} \ ;$$

then substitution yields

$$Q(\mathbf{x}(\mathbf{y})) = \mathbf{y}^t B^t C B \mathbf{y} + 2[\mathbf{b}^t C + \mathbf{c}^t]B\mathbf{y} + \mathbf{b}^t C \mathbf{b} + 2\mathbf{c}^t\mathbf{b} + c = 0 \ ,$$

which is abbreviated by

$$\overline{Q}(\mathbf{y}) = \mathbf{y}^t \overline{C}\mathbf{y} + 2\overline{\mathbf{c}}^t\mathbf{y} + \overline{c} \ .$$

If $\overline{C} = B^t C B \neq O$, this equation represents a quadric in $\mathcal{B}$. Note that $\overline{C}$ is symmetric. On the other hand, if $\overline{C} = O$ but $\overline{\mathbf{c}} \neq \mathbf{o}$, the equation is linear in $\mathbf{y}$ and represents a hyperplane in $\mathcal{B}$. If $\overline{C} = O$ and $\overline{\mathbf{c}} = \mathbf{o}$, but $\overline{c} \neq 0$, the intersection is empty; if $\overline{c} = 0$ the subspace $\mathcal{B}$ is completely contained in Q.

In the case where $\mathcal{B} = \mathcal{A}$, the intersection $\overline{Q} = \mathcal{B} \cap Q$ equals Q. Hence, $\overline{Q}(\mathbf{y}) = 0$ represents Q with respect to a different affine system. Note that $\overline{C} \neq O$, i.e., a quadric is a quadric in every affine system.

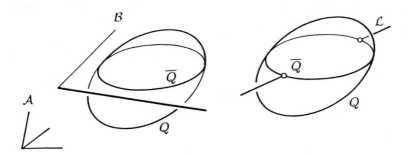

Figure 13.1: Intersection of a quadric.

In the case where $\mathcal{B}$ is an affine line $\mathcal{L}$ represented by

$$\mathbf{x} = \mathbf{b} + \mathbf{v}\lambda ,$$

the substitution above results in

$$Q(\mathbf{x}(\mathbf{y})) = \mathbf{v}^t C \mathbf{v} \lambda^2 + 2\left[C\mathbf{b} + \mathbf{c}\right]^t \mathbf{v}\lambda + \mathbf{b}^t C\mathbf{b} + 2\mathbf{c}^t\mathbf{b} + c = 0$$

or more concisely

$$\overline{Q}(\lambda) = \alpha\lambda^2 + 2\beta\lambda + \gamma = 0 .$$

If $\alpha \neq 0$, this equation represents a quadric in $\mathcal{L}$ consisting of two (real or non-real) points which can coalesce to a double point. If $\alpha = 0$ and $\beta \neq 0$, $\overline{Q}$ represents a single point on $\mathcal{L}$, and if $\alpha = \beta = 0$, $\overline{Q}$ equals the line $\mathcal{L}$ or is empty.

Remark 1: A quadric is understood to be the entire set of all real and all non-real points satisfying a quadratic equation. Consequently, equations differing by more than a factor define different quadrics.

Remark 2: A quadric is said to be **real** if all its coefficients are real. However, there need not be any real point on a quadric. In this case the quadric is called a **null quadric** or, occasionally, an **imaginary quadric**.

13.2 Midpoints

A point $\mathbf{m} = \mathbf{b}$ is called a **midpoint** of a quadric Q if Q is symmetric with respect to $\mathbf{m}$, i.e., if all straight lines $\mathbf{x} = \mathbf{m} + \mathbf{v}\lambda$ intersect Q symmetrically such that $\overline{Q}(\lambda) = 0$ implies $\overline{Q}(-\lambda) = 0$. This is the case if and only if $\beta = \mathbf{v}^t\left[C\mathbf{m} + \mathbf{c}\right] = 0$ for all $\mathbf{v}$ or, i.e., if

$$C\mathbf{m} + \mathbf{c} = \mathbf{o} .$$

The solution of $C\mathbf{m} + \mathbf{c} = \mathbf{o}$ defines an affine subspace $\mathcal{M}$ of $\mathcal{A}$, which may possibly be empty.

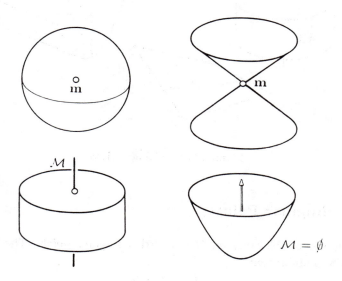

Figure 13.2: Midpoints.

Each direction $\mathbf{d}$ solving the corresponding homogeneous system $C\mathbf{d} = \mathbf{o}$ is called an **axial direction**. The axial directions span a linear space, $\mathbf{M}$. If $\mathcal{M}$ is non-empty, then $\mathbf{M}$ is the linear space underlying $\mathcal{M}$.

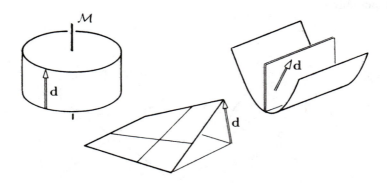

Figure 13.3: Axial directions.

13.3 Singular Points

A midpoint **s** which lies on Q is called a **singular point**. The midpoint
condition implies that

$$C\mathbf{s} + \mathbf{c} = \mathbf{o} \ ,$$

and with the additional condition $Q(\mathbf{s}) = \mathbf{s}^t\,[C\mathbf{s} + \mathbf{c}] + \mathbf{c}^t\mathbf{s} + c = 0$, it also
implies that

$$\mathbf{c}^t\mathbf{s} + c = 0 \ .$$

The solution of both equations defines an affine subspace $\mathcal{S}$ of $\mathcal{A}$, which
may possibly be empty. In particular, if $\mathcal{S} \neq \emptyset$, one has $\mathcal{S} = \mathcal{M}$.

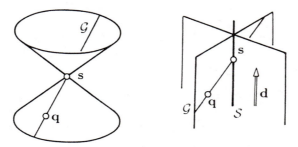

Figure 13.4: Singular points.

Remark 3: Let s be a singular point, and let q be some point of Q different from s. Then, since s is a midpoint, the line spanned by s and q intersects Q in a third point not equal to s or q, which means that this line lies completely on Q. Moreover, the join of S with some point q on Q but not in S is an affine space $\mathcal{G}$ of dimension $dim\,S + 1$. Every such $\mathcal{G}$ lies on Q, i.e., Q is a **cone** with **center** S and **generators** $\mathcal{G}$. If the center consists of only one point, it is also called **vertex**.

13.4 Tangents

A straight line $\mathcal{L}$ represented by $\mathbf{x} = \mathbf{b} + \mathbf{v}\lambda$ intersects Q in $\mathbf{q} = \mathbf{b}$ if $\lambda = 0$ is a root of $\overline{Q}(\lambda) = 0$, i.e., if $\gamma = \mathbf{q}^t C \mathbf{q} + 2\mathbf{c}^t \mathbf{q} + c = 0$. Furthermore, it is tangent to Q at $\mathbf{q}$ if $\lambda = 0$ is a double root, i.e., if

$$\beta = [C\mathbf{q} + \mathbf{c}]^t\, \mathbf{v} = 0 \;, \qquad \text{while} \;\; \alpha = \mathbf{v}^t C \mathbf{v} \neq 0 \;.$$

In this case, $\mathbf{v}$ is called **tangential** to Q at $\mathbf{q}$, and the line $\mathcal{L}$ is said to be a **tangent** of Q at $\mathbf{q}$. If in addition $\alpha = 0$, then $\mathcal{L}$ lies completely on Q; it is a generator. Note that a line through a singular point is either a tangent or a generator.

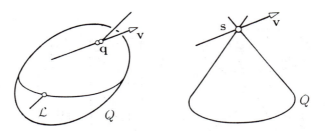

Figure 13.5: Tangents of a quadric.

13.5 Tangent Planes

Consider a point $\mathbf{q}$ on Q and some point $\mathbf{x}$ distinct from $\mathbf{q}$. The difference $\mathbf{v} = \mathbf{x} - \mathbf{q}$ is tangential to Q at $\mathbf{q}$ if

$$[C\mathbf{q} + \mathbf{c}]^t [\mathbf{x} - \mathbf{q}] = 0 .$$

Adding $Q(\mathbf{q}) = \mathbf{q}^t C\mathbf{q} + 2\mathbf{c}^t\mathbf{q} + c = 0$ to this equation one obtains

$$[C\mathbf{q} + \mathbf{c}]^t \mathbf{x} + \mathbf{c}^t\mathbf{q} + c = 0 ,$$

concisely written as $\mathbf{u}^t\mathbf{x} + u_0 = 0$. This equation, in which $\mathbf{x}$ varies, represents the equation of a hyperplane, the **tangent plane** $\mathcal{T}$ of Q at $\mathbf{q}$. Note that the tangent plane is not defined at a singular point. This property characterizes singular points, by the way.

Remark 4: A tangent plane of Q at some point $\mathbf{q}$ intersects Q in a quadric $\overline{Q}$. Since $\mathbf{q}$ is a singular point of $\overline{Q}$, the quadric $\overline{Q}$ is a cone. It may happen, though, that $\mathbf{q}$ is the only real point of $\overline{Q}$.

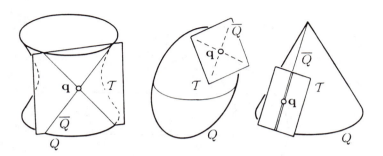

Figure 13.6: Tangent planes of a quadric.

13.6 Polar Planes

Consider a fixed point $\mathbf{x}$. It lies in the tangent plane of Q at $\mathbf{y} = \mathbf{q}$ if

$$\mathbf{x}^t C \mathbf{y} + \mathbf{c}^t (\mathbf{x} + \mathbf{y}) + c = 0 .$$

If $\mathbf{x}$ is not a singular point of Q, this is the equation of a hyperplane, the so-called **polar plane** $\mathcal{X}$ of $\mathbf{x}$ with respect to Q. The point $\mathbf{x}$ is referred to as the **pole** of $\mathcal{X}$. Pole and polar plane have the following geometric meaning. The tangent plane of Q at $\mathbf{q}$ contains $\mathbf{x}$ if and only if $\mathbf{q}$ belongs to $\mathcal{X}$, as illustrated in Figure 13.7. Note that the polar plane of some point $\mathbf{x}$ of Q coincides with the tangent plane there and is undefined if $\mathbf{x}$ is a singular point.

The equation of $\mathcal{X}$ is symmetric in $\mathbf{x}$ and $\mathbf{y}$. Consequently, if $\mathbf{y}$ lies in the polar plane $\mathcal{X}$ of $\mathbf{x}$, then $\mathbf{x}$ also lies in the polar plane $\mathcal{Y}$ of $\mathbf{y}$. Such pairs of points are called **conjugate** with respect to Q.

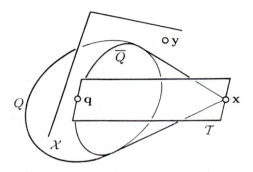

Figure 13.7: The polar plane $\mathcal{X}$ of **x**.

Example 1: One can find $n+1$ points which are pairwise conjugate with respect to a given quadric in $\mathcal{A}^n$. A general method to find such points is presented in Section 16.3. Figure 13.8 shows three pairwise conjugate points of an ellipse.

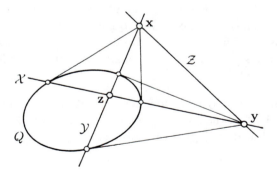

Figure 13.8: Polar triangle of an ellipse.

Remark 5: If two points $\mathbf{x}$ and $\mathbf{y}$ of a subspace $\mathcal{B}$ of $\mathcal{A}$ are conjugate with respect to a quadric Q in $\mathcal{A}$, then they are also conjugate with respect to the intersection $Q \cap \mathcal{B}$ and vice versa.

Remark 6: In particular, if $\mathcal{B}$ is a line $\mathbf{x} = \mathbf{b} + \mathbf{v}\lambda$, the points corresponding to λ_1 and λ_2 are conjugate with respect to Q if

$$\alpha \lambda_1 \lambda_2 + \beta(\lambda_1 + \lambda_2) + \gamma = 0 \ ,$$

where α, β, γ are as in Section 13.1.

13.7 Notes and Problems

1 If $\det C \neq 0$, there exists exactly one midpoint $\mathbf{m}$; it may even be singular.

2 The points on a quadric are **self-conjugate**. This property characterizes the points on a quadric.

3 A singular point is conjugate to every point. Consequently each polar plane of a quadric contains $\mathcal{S}$.

4 For any plane $\mathcal{Y}$ given by $\mathbf{u}^t \mathbf{x} + u_0 = 0$ the equations

$$C\mathbf{y} + \mathbf{c} = \mathbf{u}\varrho \ , \qquad \mathbf{c}^t \mathbf{y} + c = u_0 \varrho$$

represent a linear system for the pole $\mathbf{y}$ and ϱ.

5 The linear system of Note 4 has a unique solution if and only if

$$\det \begin{bmatrix} C & \mathbf{u} \\ \mathbf{c}^t & u_0 \end{bmatrix} \neq 0 \ .$$

6 The **interior** of a quadric Q consists of all points not on Q from which one cannot draw any real tangent to Q. The **exterior** is formed by all other points not on Q.

7 Figure 13.8 also illustrates how one can find the polar $\mathcal{Z}$ of a point $\mathbf{z}$ inside a quadric.

8 Ellipses, parabolas, and hyperbolas have an interior.

9 The midpoint of an ellipse is an inner point, but the midpoint of a hyperbola is an exterior point.

10 An ellipsoid has an interior, but a hyperboloid of one sheet has none.

11 A quadric on a line is a real or imaginary pair of points. Its midpoint is real. The midpoint is a singular point only if the quadric degenerates to a double point.

14 More on Affine Quadrics

Affine transformations map quadrics into quadrics, which means that affinely related quadrics have the same coordinate representations with respect to particular coordinate systems. As a result, there are only a finite number of affinely different types of quadrics in $\mathcal{A}^n$. One can use the fact that midpoints, singular points, tangents, etc. are preserved under affine maps to construct coordinate systems with respect to which the equations of each type of quadric take on the same simple, so-called normal, form.

Literature: Berger, Blaschke, Samuel

14.1 Diametric Planes

The midpoints of a family of parallel chords of a quadric Q lie in a plane. In order to prove this let the chords be represented by

$$\mathbf{p} = \mathbf{b} + \mathbf{v}\lambda , \quad \text{where } \mathbf{v} \text{ is fixed and } \mathbf{b} \text{ varies.}$$

The intersection points of one of these chords with Q are found by solving

$$\alpha\lambda^2 + 2\beta\lambda + \gamma = 0 ,$$

where $\alpha = \mathbf{v}^t C\mathbf{v}, \beta = [C\mathbf{b} + \mathbf{c}]^t\mathbf{v}$, and $\gamma = Q(\mathbf{b})$. In particular, $\mathbf{b}$ is the midpoint of the chord if

$$\beta = \mathbf{v}^t C\mathbf{b} + \mathbf{c}^t\mathbf{v} = 0 ,$$

which is a linear equation for $\mathbf{b} = \mathbf{x}$, concisely denoted as $\mathbf{u}^t\mathbf{x} + u_0 = 0$ and called the **diametric plane** of Q with respect to $\mathbf{v}$. Examples are illustrated in Figure 14.1. A diametric plane $\mathcal{V}$ contains all midpoints of Q, i.e., one has $\mathcal{M} \subset \mathcal{V}$. Note that the diametric plane does not exist if $\mathbf{v}$ is axial, i.e., if $C\mathbf{v} = \mathbf{o}$.

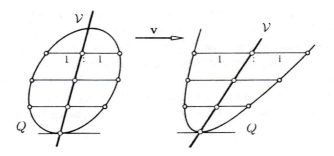

Figure 14.1: Parallel chords and diametric planes.

Obviously, the diametric plane $\mathcal{V}$ of Q with respect to $\mathbf{v}$ contains all the points of Q where $\mathbf{v}$ is tangential, as illustrated in Figure 14.2.

Remark 1: The intersection $\overline{Q}$ of $\mathcal{V}$ with Q is called the **silhouette curve** of Q in a parallel projection in the direction of $\mathbf{v}$.

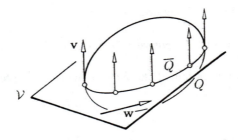

Figure 14.2: Diametric plane and tangents.

14.2 Conjugate Directions

Let $\mathbf{w}$ be the difference of two arbitrary points in the diametric plane $\mathcal{V}$. From Section 14.1 it follows that

$$\mathbf{v}^t C \mathbf{w} = 0 .$$

Two such directions $\mathbf{v}$ and $\mathbf{w}$ are called **conjugate** to each other with respect to Q. Note that this relation is symmetric in $\mathbf{v}$ and $\mathbf{w}$.

In particular, if $\mathbf{v}$ is **self conjugate**, i.e., if

$$\alpha = \mathbf{v}^t C \mathbf{v} = 0 ,$$

$\mathbf{v}$ is called **asymptotic**. From Section 13.1 it follows that any line with an asymptotic direction can either have none, one, or all points in common with Q but can never have exactly two (distinct or coalescing) points in common with Q.

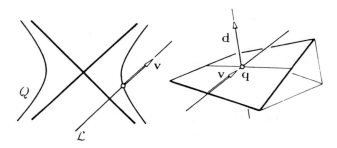

Figure 14.3: Lines with asymptotic directions.

Recall from Section 13.2 and Figure 13.3 that a direction $\mathbf{v}$ conjugate to all directions is an **axial direction** and vice versa, and that an axial direction is **singular** if, in addition, $\mathbf{c}^t \mathbf{v} = 0$.

Remark 2: If $\mathcal{M} = \emptyset$, then there exists a non-singular axial direction $\mathbf{v}$.

Remark 3: The direction of a generator, i.e., of a line lying completely on Q, is **asymptotic**. Moreover, if Q is a cylinder, the directions of its generators are also axial and singular.

Remark 4: Asymptotic directions can be regarded as the ideal points of an affine quadric (see Section 23.1).

14.3 Special Affine Coordinates

Each zero coefficient of the quadratic equation

$$Q(\mathbf{x}) = \mathbf{x}^t C \mathbf{x} + 2\mathbf{c}^t \mathbf{x} + c = 0, \quad C = C^t = [\mathbf{c}_1 \ldots \mathbf{c}_n] \ ,$$

corresponds uniquely to a special position of Q relative to the affine coordinate system.

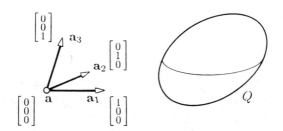

Figure 14.4: Affine system and quadric.

Let $\mathbf{a}; \mathbf{a}_1, \ldots, \mathbf{a}_n$ be the affine system. Recall that $\mathbf{o}; \mathbf{i}_1, \ldots, \mathbf{i}_n$ represents the affine system with respect to itself, where $[\mathbf{i}_1 \ldots \mathbf{i}_n]$ is the identity matrix. Thus, one can observe the following facts:

1 The **origin** lies on Q for $c = 0$.

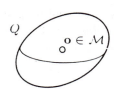

2 The origin is a **midpoint** of Q for $\mathbf{c} = \mathbf{o}$.

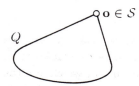

3 The origin is a **singular** point of Q for $\mathbf{c} = \mathbf{o}$ and $c = 0$.

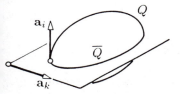

4 The directions $\mathbf{a}_i$ and $\mathbf{a}_k$ are **conjugate** with respect to Q for $c_{i,k} = 0$.

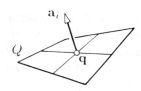

5 In particular, $\mathbf{a}_i$ is **asymptotic** for $c_{i,i} = 0$.

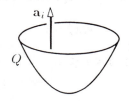

6 Furthermore, $\mathbf{a}_i$ is axial for $\mathbf{c}_i = \mathbf{o}$, where $\mathbf{c}_i$ is the ith column of C, and it is **singular** if additionally $c_i = 0$.

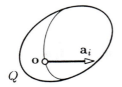

7 The direction $\mathbf{a}_i$ is tangential to Q at the origin for $c = 0$ and $c_i = 0$.

14.4 Affine Normal Forms

If C is a diagonal matrix, i.e., $c_{i,k} = 0$ for $i \neq k$, then $\mathbf{a}_1, \ldots, \mathbf{a}_n$ are pairwise conjugate. Such a **conjugate system** can always be constructed, (see Figure 13.8). Moreover, given a quadric Q, one can always choose the affine system in such a way that the equation of Q takes on one of three **normal forms** (see Section 18.5).

As before let $\mathcal{M}$ and $\mathcal{S}$ denote the solutions of

$$C\mathbf{m} + \mathbf{c} = \mathbf{o} \quad \text{and} \quad C\mathbf{s} + \mathbf{c} = \mathbf{o}, \quad \mathbf{c}^t\mathbf{s} + c = o,$$

respectively, and let $\mathbf{M}$ and $\mathbf{S}$ denote the solutions of the corresponding homogeneous systems. Note that if $\mathcal{M} = \emptyset$, then $dim\,\mathbf{M} \geq 1$ and $dim\,\mathbf{S} = dim\,\mathbf{M} - 1$. Finally, let $r = n - dim\,\mathbf{M}$.

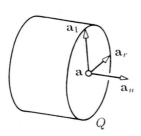

If $\mathcal{M} \neq \emptyset$, but $\mathcal{S} = \emptyset$, the quadric is called a **central quadric**. If one chooses a midpoint as the origin of the affine system, and all the $\mathbf{a}_i$ pairwise conjugate with $\mathbf{a}_{r+1}, \ldots, \mathbf{a}_n$ spanning $\mathbf{M}$, then the quadric, after an appropriate scaling, is represented by

$$\varepsilon_1 x_1^2 + \cdots + \varepsilon_r x_r^2 = 1,$$

where $\varepsilon_i = \pm 1$, $1 \leq r \leq n$.

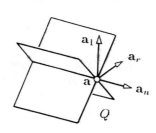

If $\mathcal{S} \neq \emptyset$, the quadric is called a **cone**. If one chooses a singular point as the origin of the affine system, and all the $\mathbf{a}_i$ pairwise conjugate with $\mathbf{a}_{r+1}, \ldots, \mathbf{a}_n$ spanning $\mathbf{S}$, then the quadric, after an appropriate scaling, is represented by

$$x_1^2 + \varepsilon_2 x_2^2 + \cdots + \varepsilon_r x_r^2 = 0 ,$$

where $\varepsilon_i = \pm 1$, $1 \leq r \leq n$.

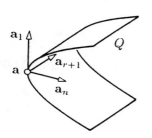

If $\mathcal{M} = \emptyset$, the quadric is called a **paraboloid**. If one chooses a point of Q as the origin of the affine system, and all the $\mathbf{a}_i$ pairwise conjugate with $\mathbf{a}_{r+1}, \ldots, \mathbf{a}_n$ spanning $\mathbf{M}$, $\mathbf{a}_n \notin \mathbf{S}$, and $\mathbf{a}_1, \ldots, \mathbf{a}_r$ tangential to Q at $\mathbf{o}$, then the quadric is represented by

$$x_1^2 + \varepsilon_2 x_2^2 + \cdots + \varepsilon_r x_r^2 = 2x_n ,$$

where $\varepsilon_i = \pm 1$, $1 \leq r < n$.

Remark 5: A quadric with singular axial directions is called a **cylinder**.

Remark 6: The affine normal form of a quadric is unique up to a permutation of the coordinates. The respective special affine system, however, is not uniquely defined.

Remark 7: A special affine system can be constructed using the Euclidean principal axes transformation described in Section 18.5.

Remark 8: Midpoints and singular points are affine invariants which classify affine quadrics. This means, e.g., that a central quadric cannot be the affine image of a paraboloid or a cone. Moreover, a central quadric can further be classified by the numbers of positive and negative ε_i. These numbers, which are given by the numbers of real and non-real intersections with the coordinate axes of the special system, are affinely invariant. This classification reflects **Sylvester's theorem**. In particular, the quadric

$$x_1^2 + \cdots + x_n^2 = 1$$

is called an **ellipsoid**. Similarly, one can classify paraboloids and cones by the numbers of positive and negative ε_i. The resulting types in $\mathcal{A}^2$ and $\mathcal{A}^3$ are presented in the next two sections.

14.5 The Types of Quadrics in the Plane

The quadrics of an affine plane are called **conics** or **conic sections**. Due to their different normal forms there are 9 different types of conic sections. Six of them contain more than one real point. They are shown in Figure 14.6.

The other three types of conics containing no or only one real point are:

the imaginary central conic section $-x_1^2 - x_2^2 = 1,$

the imaginary pair of intersecting lines $x_1^2 + x_2^2 = 0,$ and

the imaginary pair of parallel lines $-x_1^2 = 1.$

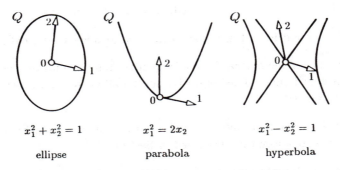

$$x_1^2 + x_2^2 = 1 \qquad\qquad x_1^2 = 2x_2 \qquad\qquad x_1^2 - x_2^2 = 1$$

ellipse   parabola   hyperbola

Figure 14.5 a: Non-degenerate conic sections.

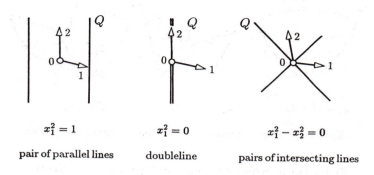

$$x_1^2 = 1 \qquad\qquad x_1^2 = 0 \qquad\qquad x_1^2 - x_2^2 = 0$$

pair of parallel lines   doubleline   pairs of intersecting lines

Figure 14.5 b: Reducible conic sections.

14.6 The Types of Quadrics in Space

Due to their different normal forms there are 17 different types of quadrics in $\mathcal{A}^3$. One can say that 12 of them contain more real points than a line. They are shown in Figure 14.6.

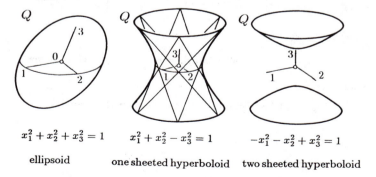

$$x_1^2 + x_2^2 + x_3^2 = 1 \qquad x_1^2 + x_2^2 - x_3^2 = 1 \qquad -x_1^2 - x_2^2 + x_3^2 = 1$$

ellipsoid   one sheeted hyperboloid   two sheeted hyperboloid

Figure 14.6 a: Non-degenerate central quadrics.

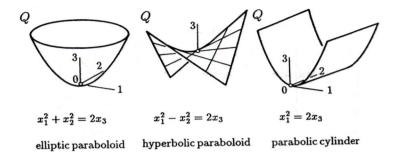

$$x_1^2 + x_2^2 = 2x_3 \qquad\qquad x_1^2 - x_2^2 = 2x_3 \qquad\qquad x_1^2 = 2x_3$$

elliptic paraboloid   hyperbolic paraboloid   parabolic cylinder

Figure 14.6 b: Non-degenerate and degenerate paraboloids.

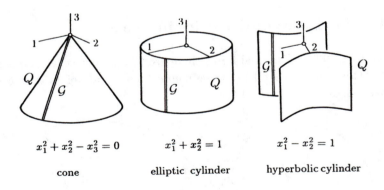

$$x_1^2 + x_2^2 - x_3^2 = 0 \qquad\qquad x_1^2 + x_2^2 = 1 \qquad\qquad x_1^2 - x_2^2 = 1$$

cone   elliptic cylinder   hyperbolic cylinder

Figure 14.6 c: Cone and degenerate central quadrics.

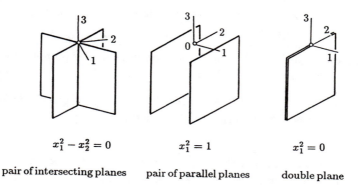

$$x_1^2 - x_2^2 = 0 \qquad\qquad x_1^2 = 1 \qquad\qquad x_1^2 = 0$$

pair of intersecting planes pair of parallel planes double plane

Figure 14.6 d: Reducible quadrics.

The other 5 types of quadrics are:

the imaginary central quadric $-x_1^2 - x_2^2 - x_3^2 = 1$,

the imaginary cylinder $-x_1^2 - x_2^2 = 1$,

the imaginary pair of parallel planes $-x_1^2 = 1$,

the imaginary cone $x_1^2 + x_2^2 + x_3^2 = 0$,

the imaginary pair of intersecting planes $x_1^2 + x_2^2 = 0$.

14.7 Notes and Problems

1 A quadric is said to be **degenerate** if it contains a singular point or a singular axial direction, i.e., if it is a cone or a cylinder.

2 A quadric is **reducible** if its equation is reducible, i.e., if it can be written as the product of two linear equations. A reducible quadric consists of a pair of hyperplanes.

3 Each non-degenerate quadric in 3-space has two families of generators such that there are two generators through each point. The tangent plane at a point is spanned by the generators through this point.

4 If the two families of lines in Note 3 are real, the quadric is called **annular** and otherwise it is called **oval**.

5 The generators of a one-sheet hyperboloid and of a hyperbolic paraboloid are real. Therefore these quadrics are called **ruled**.

6 A non-reducible but degenerate quadric in 3-space has two coalescing families of generatrices.

7 The name "cone" for quadrics which contain singular points comes from the property mentioned in Remark 2 in Section 13.3.

8 The condition $\overline{C} = B^t C B \neq 0$ in Section 13.1 means geometrically that there is at least one direction of $\mathcal{U}$ whose diametric plane is not parallel to $\mathcal{U}$.

9 An imaginary quadric can be transformed into a real one by an imaginary affine map and vice versa, see Section 23.7.

10 The equation of a quadric in $\mathcal{A}^n$ depends linearly on $\frac{1}{2}n(n+1) + n + 1$ homogeneous coefficients.

11 As a consequence of Note 10 a conic section in general position is defined by 5 points.

12 A quadric in general position in space is defined by 9 points.

13 The equation of a conic section defined by the 5 points $\mathbf{x}_1, \ldots, \mathbf{x}_5$ can be written as

$$
\det \begin{bmatrix}
x^2 & y^2 & x\,y & x & y & 1 \\
x_1^2 & y_1^2 & x_1 y_1 & x_1 & y_1 & 1 \\
\vdots & & \vdots & & & \vdots \\
x_5^2 & y_5^2 & x_5 y_5 & x_5 & y_5 & 1
\end{bmatrix} = 0 \; .
$$

15 Homothetic Pencils

Many properties of quadrics do not depend on the constant terms of their equations. These properties reveal interesting relations among quadrics that differ only in the constant terms of their equations. Such a family of quadrics is called a homothetic pencil, and the family's definition does not depend on a specific affine coordinate system. In particular, homothetic pencils are useful in analyzing intersections of quadrics with pencils of parallel lines and planes.

Literature: Berger, Coxeter, Samuel

15.1 The Equation

The family of quadrics represented by

$$Q(\mathbf{x}, c) = \mathbf{x}^t C \mathbf{x} + 2\mathbf{c}^t \mathbf{x} + c = 0 \ ,$$

where C and $\mathbf{c}$ are fixed, but c varies is called a **homothetic pencil**. Since $Q(\mathbf{x}, c)$ is linear in c, every point $\mathbf{x}$ lies on exactly one member of the pencil. This property has two immediate consequences. The quadrics of such a pencil are pairwise disjoint, and the pencil covers the entire space. Moreover, because of Remark 1 in Section 13.1, a homothetic pencil is determined by any one of its members.

It follows from Chapter 13 that the definition of a homothetic pencil does not depend on a particular affine system. Moreover, all quadrics of a

homothetic pencil have the same midpoints, if there are any; the same
diametric plane with respect to some arbitrary direction, and therefore
the same pairs of conjugate directions; and if there are any, the same
asymptotic and the same axial directions.

Furthermore, the polar planes of some fixed pole **x** with respect to the
quadrics of a homothetic pencil form a pencil of parallel planes. One of
these planes is the tangent plane at **x** of the unique quadric containing **x**.

Note that a homothetic pencil can also contain imaginary quadrics.

15.2 Asymptotic Cones

Consider a homothetic pencil of quadrics with a midpoint **m**, and let Q_0 be
the unique quadric of the pencil containing **m**. According to Section 13.3
the midpoint **m** is a singular point of Q_0. Hence, Q_0 is a cone, called the
asymptotic cone of all quadrics of the pencil. Note that the points of $\mathcal{M}$
might be the only real points of Q_0. Note also that paraboloids do not
possess an asymptotic cone.

Example 1: On using the normal form, a homothetic pencil of central
quadrics is represented by

$$\varepsilon_1 x_1^2 + \cdots + \varepsilon_r x_r^2 = c .$$

Figure 15.1 shows the corresponding three types of homothetic pencils in
the plane.

Remark 1: An important property of homothetic central quadrics is that
they are centric affine with respect to every $\mathbf{m} \in \mathcal{M}$. More exactly, the
centric affinity $\mathbf{y} = \mathbf{x}\varrho$ transforms the quadric given by $\varepsilon_1 x_1^2 + \cdots + \varepsilon_r x_r^2 = c$
into the one given by

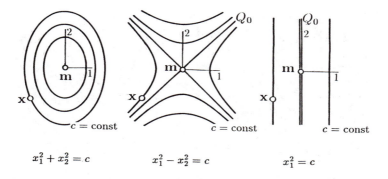

Figure 15.1: Homothetic pencils of central conic sections.

$$\varepsilon_1 y_1^2 + \cdots + \varepsilon_r y_r^2 = c\varrho^2 .$$

This property accounts for the name **homothetic**. Each such centric affinity maps the asymptotic cone onto itself. Note that the three hyperbolas in Figure 15.1 are real members of the same pencil. Moreover, a one-sheet hyperboloid is always homothethic to a hyperboloid of two sheets having the same asymptotic cone. They are related by an imaginary dilatation.

Example 2: A homothetic pencil of quadrics on an affine line consists of pairs of points which are symmetric with respect to a point **m**.

15.3 Homothetic Paraboloids

If the quadrics of a homothetic pencil are paraboloids, they have no midpoint and therefore no asymptotic cone. However, analogous to the centric affinity above, one has the important property that homothetic paraboloids are translates of each other. Namely, if the pencil is given in normal form,

$$\varepsilon_1 x_1^2 + \cdots + \varepsilon_r x_r^2 = 2x_n + c ,$$

one can observe that the members are simple translates of each other in the direction of x_n. The direction of x_n represents a non-singular axial direction **d** as mentioned in Section 14.4. Figure 15.2 shows the three types of homothetic paraboloids in $\mathcal{A}^3$.

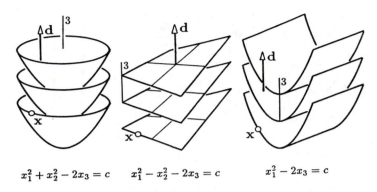

$$x_1^2 + x_2^2 - 2x_3 = c \qquad x_1^2 - x_2^2 - 2x_3 = c \qquad x_1^2 - 2x_3 = c$$

Figure 15.2: Homothetic paraboloids.

15.4 Intersection with a Subspace

According to Section 13.1 the intersections of homothetic quadrics with an affine subspace $\mathcal{B}$ either forms another homothetic pencil of quadrics or a pencil of parallel subspaces, unless $\mathcal{B}$ lies in one of the quadrics itself. The dimension and the position of $\mathcal{B}$ determine which of these cases one actually has.

Any non-asymptotic line intersects a pencil of quadrics in pairs of points symmetric with respect to some fixed point $\mathbf{n}$ and, consequently, is tangent to the quadric of the pencil containing $\mathbf{n}$. This symmetry property leads to a simple construction of the quadric Q_1 containing $\mathbf{p}$ and homothetic to some given quadric Q: Every non-asymptotic line $\mathbf{p} + \mathbf{v}\lambda$ intersects Q in two points $\mathbf{p} + \mathbf{v}\lambda_1$ and $\mathbf{p} + \mathbf{v}\lambda_2$ which could be imaginary or identical. In any case, the point

$$\mathbf{x} = \mathbf{p} + \mathbf{v}(\lambda_1 + \lambda_2)$$

lies on Q_1, and if $\mathbf{v}$ varies, $\mathbf{x}$ traces out the entire quadric Q_1. Note that $\lambda_1 + \lambda_2$ is always real, although λ_1 and λ_2 may be complex.

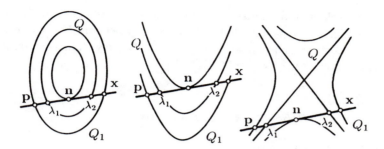

Figure 15.3: Intersection with a straight line.

Example 3: Consider a one-sheet hyperboloid Q and its asymptotic cone Q_0. Each tangent plane T_0 of Q_0 intersects Q_0 in a double line G_0. Consequently, T_0 intersects Q in two lines G_1 and G_2 parallel to G_0. If T_0 varies on Q_0, both lines G_1 and G_2 vary on Q forming the two **families of generators** of the one-sheet hyperboloid, as illustrated in Figure 15.4

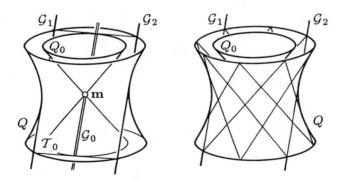

Figure 15.4: Asymptotic cone and generators of a hyperboloid of one sheet.

15.5 Parallel Intersections

The intersections of a quadric with two parallel subspaces are related by
means of homothetic pencils. Let $\mathcal{B}$ be the subspace defined by $\mathbf{x} = \mathbf{b} + B\mathbf{y}$,
and let $\mathcal{B}^*$ be its translation defined by $\mathbf{x} = \mathbf{b} + \triangle\mathbf{b} + B\mathbf{y}$. Their intersections
with the quadric Q, $Q(\mathbf{x}) = \mathbf{x}^t C\mathbf{x} + 2\mathbf{c}^t\mathbf{x} + c$, are given by

$$\overline{Q} = Q(\mathbf{x}(\mathbf{y}))$$
$$= \mathbf{y}^t B^t C B\mathbf{y} + 2\left[C\mathbf{b} + \mathbf{c}\right]^t B\mathbf{y} + Q(\mathbf{b}) = 0 \ ,$$

$$\overline{\overline{Q}} = Q(\mathbf{x}(\mathbf{y}) + \triangle\mathbf{b})$$
$$= \mathbf{y}^t B^t C B\mathbf{y} + 2\left[C\left[\mathbf{b} + \triangle\mathbf{b}\right] + \mathbf{c}\right]^t B\mathbf{y} + Q(\mathbf{b} + \triangle\mathbf{b}) = 0 \ ,$$

respectively. $\overline{Q}$ and $\overline{\overline{Q}}$ each define a homothetic pencil. These pencils are
identical up to a translation if $B^t C \triangle\mathbf{b} = \mathbf{o}$, i.e., if $\triangle\mathbf{b}$ is conjugate to all
directions of $\mathcal{B}$ with respect to Q.

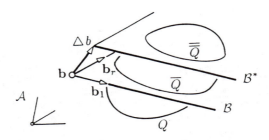

Figure 15.5: Parallel intersections of a quadric.

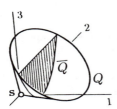

Example 4: Consider the cone Q
given by the equation

$$2x_1 x_3 = x_2^2 \ .$$

The pencil of parallel planes $x_1 =$
const intersects Q in a family of conic
sections which are central affine, but

the corresponding pencils are not translates of each other since the x_1-direction and the x_3-direction are not conjugate with respect to Q.

Example 5: Consider an ellipsoid Q given by the equation

$$\frac{x^2}{a^2} + \frac{y^2}{b^2} + \frac{z^2}{c^2} = 1 ,$$

where $0 < c < b < a$ and its intersection with the sphere

$$x^2 + y^2 + z^2 = r^2 .$$

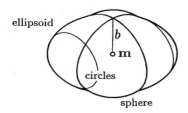

If $r = b$, the intersection consists of two circles each of which lies in a plane. The planes parallel to each of these planes also intersect Q in circles. These planes are therefore called **planes of circular sections**.

Example 6: Consider two parabolas in different planes but with a common point **p** and the same axis direction. Sweeping one parabola parallel in space such that **p** moves along the second parabola generates a paraboloid. One obtains the same paraboloid when the second parabola is swept along the first one, as illustrated to the left.

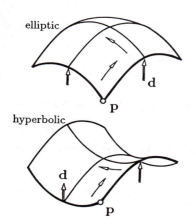

Remark 2: Planes of circular sections exist for most quadrics. The four points of a quadric where the circles degenerate to a point are the **circular** or **umbilical** points of the quadric. Note that one can distinguish spheres and circles from ellipsoids and ellipses only in a Euclidean space.

15.6 Notes and Problems

1 A polynomial of total degree 2 over a triangle represents a segment of a paraboloid if and only if the axes of the three boundary parabolas are parallel.

2 A paraboloid in 3D is defined by three pairwise intersecting parabolas having the same axis direction.

3 The volume bounded by a quadric and a tangent plane of a homothetic quadric does not depend on the position of the tangent plane, as illustrated in Figure 15.6.

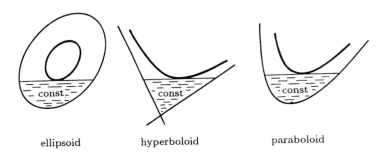

ellipsoid   hyperboloid   paraboloid

Figure 15.6 : Constant volumes.

4 The property mentioned in Note 3 is applied in shipbuilding to determine the waterline of a tilted ship.

5 Parallel intersections of a quadric occur in scan-line algorithms for computer graphics.

6 In Example 5 one also gets planes of circular sections for $r = a$ and $r = c$. However, these planes are not real.

7 Consider the one-sheet hyperboloid given by

$$\frac{x^2}{a^2} + \frac{y^2}{b^2} - \frac{z^2}{c^2} = 1, \quad b < a \ .$$

Its intersection with the sphere $x^2 + y^2 + z^2 = a^2$ consists of two circles.

8 A one-sheet hyperboloid has two pencils of circular intersections, yet it does not possess real umbilical points.

9 A homothetic pencil of quadrics can be viewed as the generalization of a pencil of parallel planes.

PART FOUR

Euclidean Geometry

As far as we know, geometry originated from land measurements and area and volume computations made by the early Babylonians and Egyptians in the years approximately from 4000 to 1500 B.C. The word *geometry* itself is derived from the Greek word for "earth measure." The Greeks (600 to 300 B.C.) developed the empirical geometry of the Babylonians and Egyptians into a more systematic science which ultimately prepared the ground for Euclid's outstanding *Elements*. Euclid, who lived in Alexandria in about 300 B.C., gave a systematic logical organization of what is now called Euclidean geometry. Euclid's work was so solid that it took 2000 years before geometries other than Euclidean geometry were discovered.

The structure of an affine space does not encompass distances and angles. If distances and angles are defined in an affine space, one refers to it as a Euclidean space. In general, affine maps do not preserve distances and angles. Those that do are the Euclidean motions.

16 The Euclidean Space

The Euclidean space is an affine space where, in addition, the distance between two points is defined in the usual way. This distance between points induces a compatible length of vectors in the underlying linear space and facilitates the introduction of angles. However, a Euclidean space can be introduced more geometrically by means of a gauge ellipsoid.

Literature: Berger, Perdoe, Coxeter

16.1 The Distance of Points

One can introduce a distance into an affine space by means of a distinguished ellipsoid, the so-called **gauge quadric**. Let G be such a gauge quadric and $\mathbf{m}$ its midpoint. Then the **distance** of two points $\mathbf{p}$ and $\mathbf{q}$ is defined as the non-negative number δ such that

$$[\mathbf{x} - \mathbf{m}]\delta = \mathbf{p} - \mathbf{q},$$

where $\mathbf{x}$ is a point of G.

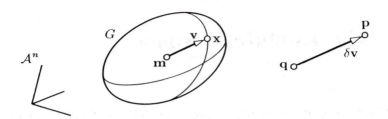

Figure 16.1: Gauge ellipsoid and Euclidean distance.

For the sake of simplicity one may assume that $\mathbf{m}$ is the origin of the affine coordinate system. Then the equation of G can be brought into the form $\mathbf{x}^t C \mathbf{x} = 1$ and one has

$$\delta^2 = [\mathbf{p} - \mathbf{q}]^t \, C \, [\mathbf{p} - \mathbf{q}] \ .$$

An affine space with such a distance is called a **Euclidean space**. In particular, the notation $\mathcal{E}^n$ means the affine space $\mathcal{A}^n$ with such a distance.

Remark 1: The gauge quadric is regarded as the unit sphere in $\mathcal{E}^n$. Hence, the members of the homothetic pencil

$$\mathbf{x}^t C \mathbf{x} = c$$

are **spheres** of radii $\varrho = \sqrt{c}$ and center $\mathbf{o}$.

16.2 The Dot Product

In a Euclidean space, two directions $\mathbf{v}$ and $\mathbf{w}$ are called **orthogonal** if they are conjugate with respect to the gauge ellipsoid. Furthermore, the **scalar** or **dot product** of two vectors $\mathbf{v}$ and $\mathbf{w}$ is defined by

$$\mathbf{v} \cdot \mathbf{w} = \mathbf{v}^t C \mathbf{w} \ ,$$

where $\mathbf{x}^t C \mathbf{x} = 1$ is the equation of the gauge quadric. The dot product is symmetric and bilinear, i.e., linear in each of its two arguments.

The underlying vector space of $\mathcal{E}^n$ is called a **Euclidean vector space** if it is associated with such a dot product. This Euclidean vector space is denoted by $\mathbf{E}^n$.

The **Euclidean length** or **norm** of a vector $\mathbf{v} \in \mathbf{E}^n$ is defined by

$$|\mathbf{v}| = \sqrt{\mathbf{v} \cdot \mathbf{v}} ,$$

which is also the distance between any two points $\mathbf{p}$ and $\mathbf{p} + \mathbf{v}$.

Multiplying a non-zero vector $\mathbf{v}$ by $1/|\mathbf{v}|$ produces a **normalized** vector of length 1.

Two vectors $\mathbf{u}$ and $\mathbf{v}$ are called **orthogonal** if $\mathbf{u} \cdot \mathbf{v} = 0$. Moreover, a family of pairwise orthogonal vectors is called **orthonormal** if in addition all vectors have unit length. Note that an orthogonal set of non-zero vectors is linearly independent.

The dot product can be used to introduce angles between two vectors $\mathbf{v}$ and $\mathbf{w}$. Let ϱ be such that $\mathbf{v} - \mathbf{w}\varrho$ is orthogonal to $\mathbf{w}$. Then the angle φ between $\mathbf{v}$ and $\mathbf{w}$ is defined by $\varrho|\mathbf{w}| = |\mathbf{v}|\cos\varphi$, as in elementary trigonometry; see Figure 16.2. Since $[\mathbf{v} - \mathbf{w}\varrho] \cdot \mathbf{w} = 0$, one has

$$\mathbf{v} \cdot \mathbf{w} = |\mathbf{v}| \cdot |\mathbf{w}| \cos\varphi ,$$

which comprises the **law of cosines**

$$|\mathbf{v} - \mathbf{w}|^2 = |\mathbf{v}|^2 - 2\mathbf{v} \cdot \mathbf{w} + |\mathbf{w}|^2 = |\mathbf{v}|^2 + |\mathbf{w}|^2 - 2|\mathbf{v}||\mathbf{w}|\cos\varphi$$

and, in particular, for $\varphi = 90°$ the **theorem of Pythagoras**.

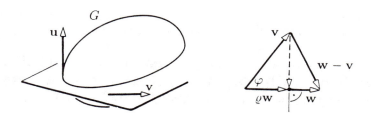

Figure 16.2: Orthogonality and dot product.

The angle φ given by three points $\mathbf{a} = \mathbf{b} + \mathbf{v}$, $\mathbf{b}$, and $\mathbf{c} = \mathbf{b} + \mathbf{w}$ is defined as the angle between the vectors $\mathbf{v}$ and $\mathbf{w}$ and denoted by

$$\not\!\!\!\times \mathbf{a}\,\mathbf{b}\,\mathbf{c} = \varphi \ .$$

16.3 Gram-Schmidt Orthogonalization

Given a basis $\mathbf{a}_1, \ldots, \mathbf{a}_r$ of some r-dimensional Euclidean vector space, one can construct an orthogonal basis $\mathbf{b}_1, \ldots, \mathbf{b}_r$ from it by what is known as the **Gram-Schmidt orthogonalization**: Starting with $\mathbf{b}_1 = \mathbf{a}_1$ one constructs $\mathbf{b}_2, \ldots, \mathbf{b}_r$ successively by adding a linear combination of $\mathbf{b}_1, \ldots, \mathbf{b}_{k-1}$ to $\mathbf{a}_k$ such that the sum $\mathbf{b}_k$ is orthogonal to $\mathbf{b}_1, \ldots, \mathbf{b}_{k-1}$. A detailed algorithmic formulation is:

for $k = 1, \ldots, r$ **do** : $\mathbf{b}_k = \mathbf{a}_k - u_{1,k}\mathbf{b}_1 - \cdots - u_{k-1,k}\mathbf{b}_{k-1}$,

where $u_{i,k} = (\mathbf{b}_i \cdot \mathbf{a}_k)/(\mathbf{b}_i \cdot \mathbf{b}_i)$.

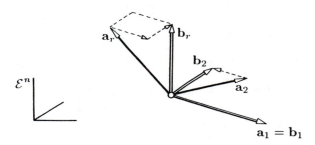

Figure 16.3: Schmidt's orthogonalization.

Usually the vectors $\mathbf{b}_k$ are normalized right after their computation, i.e., they are replaced by $\mathbf{b}_k/|\mathbf{b}_k|$. This simplifies $u_{i,k}$ to $\mathbf{b}_i \cdot \mathbf{a}_k$ and represents the common Gram-Schmidt **orthonormalization** process. In matrix notation the Gram-Schmidt process is written as

$$[\mathbf{a}_1 \ldots \mathbf{a}_r] = [\mathbf{b}_1 \ldots \mathbf{b}_r] \begin{bmatrix} u_{11} & \cdots & u_{1r} \\ & \ddots & \vdots \\ & & u_{rr} \end{bmatrix} ,$$

which represents an orthogonalization if $u_{11} = \cdots = u_{rr} = 1$ but an orthonormalization if $\mathbf{b}_i^t \mathbf{b}_i = 1$.

Remark 2: On using an arbitrary quadric Q as gauge quadric, the Gram-Schmidt orthogonalization can be used to construct a family of vectors which are pairwise conjugate to Q. However, precautions are necessary if Q has real asymptotic directions.

Remark 3: The Gram-Schmidt orthogonalization does not provide a numerically stable algorithm. Therefore, it is better to employ Householder's orthogonalization, which is described in most texts on Numerical Analysis.

16.4 Cartesian Coordinates

An affine coordinate system of $\mathcal{E}^n$ is called **Cartesian** if the equation of the gauge quadric takes on its normal form $x_1^2 + \cdots + x_n^2 = 1$. Then the dot product is simply

$$\mathbf{v} \cdot \mathbf{w} = \mathbf{v}^t \mathbf{w} = \mathbf{w}^t \mathbf{v} .$$

The basis vectors of a Cartesian system are obviously orthonormal since they are represented by the columns of the identity matrix. The simple form of the dot product has various implications:

The directions $\triangle\mathbf{x}$, i.e., the differences of points, of a hyperplane $\mathcal{U}$ given by the equation

$$\mathbf{u}^t \mathbf{x} + u_0 = 0$$

satisfy

$$\mathbf{u}^t \triangle\mathbf{x} = 0 .$$

Therefore the vector represented by the Cartesian coordinates $\mathbf{u}$ is orthogonal to the direction of $\mathcal{U}$. One says that $\mathbf{u}$ represents the **normal direction** of $\mathcal{U}$.

Let $vol_r[\mathbf{a}_1 \ldots \mathbf{a}_r]$ denote the r-dimensional volume of the parallelepiped spanned by $\mathbf{a}_1, \ldots, \mathbf{a}_r$. From the Gram-Schmidt orthonormalization one can observe that

$$vol_r[\mathbf{a}_1 \ldots \mathbf{a}_r] = u_{11} \cdots u_{rr} \ ,$$

and

$$vol_r^2[\mathbf{a}_1 \ldots \mathbf{a}_r] = \det A^t A \ ,$$

where $A = [\mathbf{a}_1 \ldots \mathbf{a}_r]$. In particular, for $r = n$ one has

$$vol_n[\mathbf{a}_1 \ldots \mathbf{a}_n] = \det A \ .$$

16.5 The Alternating Product

The expansion of the determinant of an $n \times n$ matrix $A = [\mathbf{a}_1 \ldots \mathbf{a}_n]$ along its last column can be written as

$$\det A = \det [\mathbf{a}_1 \ldots \mathbf{a}_n] = \mathbf{v}^t \mathbf{a}_n \ ,$$

where the components of $\mathbf{v}$ are

$$v_i = (-1)^{i+n} \det A_{i,n} \ ,$$

and $A_{i,n}$ is obtained from A by deleting its ith row and nth column. The vector $\mathbf{v}$ is called the **alternating product** of $\mathbf{a}_1, \ldots, \mathbf{a}_{n-1}$ and is written as

$$\mathbf{v} = \mathbf{a}_1 \wedge \cdots \wedge \mathbf{a}_{n-1} \ .$$

The alternating product has two interesting geometric properties:

1) The vector $\mathbf{v}$ is orthogonal to $\mathbf{a}_1, \ldots, \mathbf{a}_{n-1}$.

This follows from the fact that a determinant vanishes if two columns coincide.

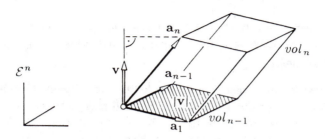

Figure 16.4: The alternating product.

2) The length of $\mathbf{v}$ equals $vol_{n-1}[\mathbf{a}_1 \ \ldots \ \mathbf{a}_{n-1}]$.

To verify this property, let $\mathbf{b}_n$ denote the unit vector parallel to $\mathbf{v}$. Then

$$vol_n[\mathbf{a}_1 \ \ldots \ \mathbf{a}_n] = |\mathbf{v}| \, \mathbf{b}_n^t \mathbf{a}_n .$$

On the other hand one can observe from Figure 16.4

$$vol_n[\mathbf{a}_1 \ \ldots \ \mathbf{a}_n] = vol_{n-1}[\mathbf{a}_1 \ \ldots \ \mathbf{a}_{n-1}]\mathbf{b}_n^t \mathbf{a}_n$$

which establishes the above claim.

Example 1: If $n = 3$, the alternating product coincides with the usual **vector product** of two vectors $\mathbf{a}_1$ and $\mathbf{a}_2$ in $\mathcal{E}^3$, i.e., $\mathbf{v}$ is orthogonal to $\mathbf{a}_1, \mathbf{a}_2$ and its length equals the area of the parallelogram spanned by $\mathbf{a}_1, \mathbf{a}_2$.

16.6 Euclidean Motions

An affine map which maps a Cartesian system onto another Cartesian system is called a **Euclidean motion**. Such maps were already used in Chapter 5. Euclidean motions clearly do not change Euclidean distances and angles. Let

$$\mathbf{x} = \mathbf{b} + B\mathbf{y}$$

describe a Euclidean motion $\mathcal{A} \rightarrow \mathcal{B}$. As was shown for affine maps, $\mathbf{b}$ and the columns $\mathbf{b}_i$ of B represent the image of the Cartesian coordinate system of $\mathcal{A}$. Since the image has to be a Cartesian system again, B is orthonormal, i.e., $B^t B = I$.

Remark 4: From $B^t B = I$ it follows that $\det{}^2 B = 1$ and therefore $\det B = \pm 1$. If $\det B = 1$, the motion is called **proper** and otherwise it is called **improper**.

An **orientation** in a vector space is defined as follows. One assigns the orientation $+1$ to some basis $\mathbf{a}_1, \ldots, \mathbf{a}_n$ and then assigns a sequence of n linearly independent vectors $\mathbf{b}_1, \ldots, \mathbf{b}_n$ the orientation $sign \det B$, where the ith column of B represents $\mathbf{b}_i$ with respect to $\mathbf{a}_1, \ldots, \mathbf{a}_n$.

A proper motion preserves and an improper motion reverses the orientation of any such sequence.

16.7 Shortest Distances

Let $B \subset \mathcal{E}^n$ be a subspace. The perpendicular from a point $\mathbf{p}$ onto B intersects B at $\mathbf{f}$, the **foot** of the perpendicular. For mechanical reasons $\mathbf{f}$ is unique, and the distance d between $\mathbf{p}$ and $\mathbf{f}$ is the shortest distance between $\mathbf{p}$ and B. The computation of the foot of a perpendicular is discussed in Sections 3.1 and 3.5. For some special cases the explicit solutions are given below:

Point and Plane

Let $\mathbf{u}^t \mathbf{x} - u_0 = 0$ be the equation of some hyperplane $\mathcal{U}$. Intersecting the perpendicular $\mathbf{x} = \mathbf{p} - \lambda \mathbf{u}$ with $\mathcal{U}$ gives

$$d = \frac{|\mathbf{u}^t \mathbf{p} - u_0|}{|\mathbf{u}|} .$$

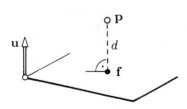

Let $\mathcal{U}$ be given by a parametric representation $\mathbf{x} = \mathbf{b} + B\mathbf{y}$ where $B = [\mathbf{b}_1 \ldots \mathbf{b}_{n-1}]$. Computing the volume spanned by $\mathbf{p} - \mathbf{b}$ and the columns of B gives

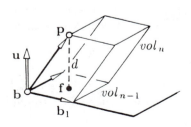

$$d = \frac{\det [\mathbf{p} - \mathbf{b} \ \mathbf{b}_1 \ldots \mathbf{b}_{n-1}]}{|\mathbf{b}_1 \wedge \cdots \wedge \mathbf{b}_{n-1}|} .$$

Note that d has a sign here.

Point and Line

Let a line $\mathcal{L}$ be given by $\mathbf{x} = \mathbf{b} + \mathbf{b}_1\lambda$. Computing the volume spanned by $\mathbf{p} - \mathbf{b}$ and $\mathbf{b}_1$ gives

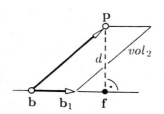

$$d = \frac{vol_2[\mathbf{p} - \mathbf{b} \ \mathbf{b}_1]}{vol_1[\mathbf{b}_1]} ,$$

which in $\mathcal{E}^3$ can be rewritten as

$$d = \frac{|(\mathbf{p} - \mathbf{b}) \wedge \mathbf{b}_1|}{|\mathbf{b}_1|} .$$

Note that $\mathbf{p}$ and $\mathcal{L}$ only span a two-dimensional space.

Line and Line

Let two lines be given by $\mathbf{x} = \mathbf{a} + \mathbf{a}_1\alpha$ and $\mathbf{x} = \mathbf{b} + \mathbf{b}_1\beta$, respectively. Computing the volume spanned by $\mathbf{a}_1, \mathbf{b}_1$ and $\mathbf{a} - \mathbf{b}$ gives

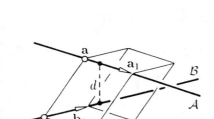

$$d = \frac{vol_3[\mathbf{a} - \mathbf{b} \ \mathbf{a}_1 \ \mathbf{b}_1]}{vol_2[\mathbf{a}_1 \ \mathbf{b}_1]} ,$$

which can be rewritten in $\mathcal{E}^3$ as

$$d = \frac{|\det[\mathbf{a} - \mathbf{b} \ \mathbf{a}_1 \ \mathbf{b}_1]|}{|\mathbf{a}_1 \wedge \mathbf{b}_1|} .$$

Remark 5: The equation $d(\mathbf{x}) = \frac{\mathbf{u}^t\mathbf{x} - u_0}{|\mathbf{u}|} = 0$ is called the **Hesse normal form** of the hyperplane $\mathcal{U}$.

16.8 The Steiner Surface in Euclidean Space

Consider a point $\mathbf{x}$ and two lines $\mathcal{A}$ and $\mathcal{B}$ with distance d. Let a and b denote the distances from $\mathbf{x}$ to $\mathcal{A}$ and $\mathcal{B}$, respectively. Then if $\mathbf{x}$ varies such that the difference or sum of a and b equals $+d$ or $-d$, $\mathbf{x}$ sweeps out a so-called parabolic **Steiner surface**.

Let $\mathcal{A}$ and $\mathcal{B}$ be given by $\mathbf{a} + \mathbf{a}_1 \alpha$ and $\mathbf{b} + \mathbf{b}_1 \beta$, respectively, then one has

$$a = \frac{|(\mathbf{x} - \mathbf{a}) \wedge \mathbf{a}_1|}{|\mathbf{a}_1|} \ , \quad b = \frac{|(\mathbf{x} - \mathbf{b}) \wedge \mathbf{b}_1|}{|\mathbf{b}_2|} \ , \quad d = \frac{\det\left[(\mathbf{a} - \mathbf{b})\mathbf{a}_1 \mathbf{b}_1\right]}{|\mathbf{a}_1 \wedge \mathbf{b}_1|} \ .$$

Substituting this into

$$(a + b + d)(a + b - d)(a - b + d)(a - b - d) = (a^2 + b^2 - d^2)^2 - 4a^2 b^2 = 0$$

gives the equation of this surface.

Example 2: Let $\mathcal{A}$ and $\mathcal{B}$ be given by

$$[c \quad 0 \quad 0]^t + [0 \quad 1 \quad 0]^t \alpha \ , \quad [-c \quad 0 \quad 0]^t + [0 \quad 0 \quad 1]^t \beta \ ,$$

respectively, as illustrated in Figure 16.5. Then $d = 2c$ and

$$a^2 = (x - c)^2 + z^2 \ , \quad b^2 = (x + c)^2 + y^2,$$

where $\mathbf{x} = [x \quad y \quad z]^t$. The resulting equation is

$$(y^2 - z^2)^2 - 8c^2(y^2 + z^2) + 8cx(y^2 - z^2) = 0 \ .$$

The intersection of the surface with any plane of the pencil $z - y \cdot \tan \varphi = 0$ degenerates to the x-axis counted twice and the parabola

$$z^2 - y^2 = 8c(x - c/\cos 2\varphi) \ , \quad z = y \cdot \tan \varphi \ .$$

Because of this fact this kind of Steiner surface is called parabolic. Note that for $\varphi = \pm 45°$ the parabola degenerates to double lines at infinity.

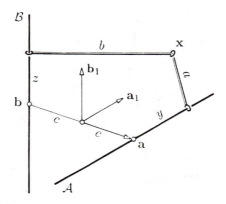

Figure 16.5: The string construction.

Remark 6: One can check by substitution that the above surface has the parameter representation in homogeneous coordinates

$$x_1 = c(u^2 + v^2 - w^2) , \quad x_2 = \sqrt{2}c \cdot 2uw , \quad x_3 = \sqrt{2}c \cdot 2vw , \quad x_0 = u^2 - v^2 .$$

Figure 16.6 shows two quadrangular patches of this rational surface (cf. Section 23.8).

Figure 16.6: The constructed Steiner surface.

16.9 Notes and Problems

1 While one can compare differences of points in an affine space only if they are parallel, one can compare differences of points in a Euclidean space independent of their direction.

2 If an hyperboloid of two sheets is used as a gauge quadric in $\mathcal{A}^3$, one gets a so-called **Minkowski world**.

3 Replacing a gauge quadric by another quadric belonging to the same homothetic pencil means that all distances are multiplied by a common factor while all angles are left unchanged.

4 An affine map of the Euclidean space which leaves angles unchanged is a **dilatation**.

5 A proper Euclidean motion which has a fixed point $\mathbf{c}$ is called a **rotation** around the **center c**.

6 Every rotation in $\mathcal{E}^3$ has a line of fixed points. This line is called its axis.

7 Two subspaces $\mathcal{A}$ and $\mathcal{B}$ are called **skew** if $\mathcal{A} \sqcap \mathcal{B} = \emptyset$ and $\mathbf{A} \sqcap \mathbf{B} = \{\mathbf{o}\}$.

8 Let $\mathcal{A}$ and $\mathcal{B}$ be two skew subspaces given by

$$\mathbf{x} = \mathbf{a} + A\mathbf{y} \quad \text{and} \quad \mathbf{x} = \mathbf{b} + B\mathbf{z} \ .$$

On computing the volume of the parallelepiped spanned by $\mathbf{a} - \mathbf{b}$ and the columns of A and B, one gets for the shortest distance between $\mathcal{A}$ and $\mathcal{B}$

$$d = \frac{vol\,[\mathbf{a} - \mathbf{b}|A|B]}{vol\,[A|B]} \ .$$

9 By definition, the zero vector is orthogonal to every vector.

17 Some Euclidean Figures

The structure of a Euclidean space is not affected by Euclidean movements. Moreover, there are properties of figures which are invariant, i.e., there are rules for constructing the figures, and the properties depend on these rules, but they do not depend on the figures' specific position in Euclidean space. Dropping a perpendicular from a point onto a plane is a simple example of such a construction.

Literature: Berger, Coxeter, Perdoe

17.1 The Orthocenter

Several classical theorems on triangles are concerned with perpendiculars from a vertex onto the opposite side:

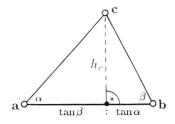

Perpendiculars
Consider a triangle **abc**. Let α, β, γ denote the corresponding angles. The **perpendicular** from **c** onto **ab** subdivides **a**, **b** in the ratio $\tan\beta : \tan\alpha$.

Orthocenter
From Ceva's theorem in Section 12.1 it follows that the three altitudes of a triangle meet in a point **d** which is

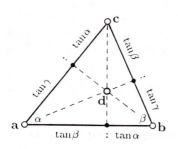

called the **orthocenter** of the triangle. Note that, for the three vertices and **d**, each of these four points is the orthocenter of the other three.

Barycentric Coordinates

It also follows from Ceva's theorem that the ratios of the barycentric coordinates of the orthocenter **d** with respect to **a**, **b**, **c** are given by $\tan \alpha : \tan \beta : \tan \gamma$.

Example 1: The last fact has been used in Section 7.4 in the discussion of the three-point perspective.

17.2 The Incircle

Other classical theorems on triangles are concerned with **bisectors**, i.e., lines which bisect the angles between two given lines.

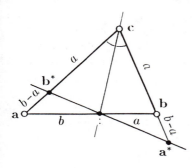

Bisector

Let a, b, and c denote the lengths of the sides opposite **a**, **b**, and **c**, respectively. By reflecting through the **bisector** of **c**, one can obtain the same situation as in Menelaus' theorem and thus can conclude the following: The bisector at **c** subdivides the opposite side in the ratio $b : a$.

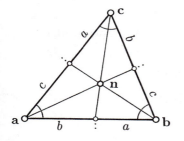

Incenter

From the fact above and Ceva's theorem, it follows that the three bisectors of a triangle meet at a point **n** whose barycentric coordinates with respect to **a**, **b**, **c** form the ratios $a : b : c$.

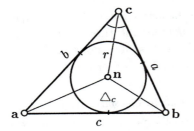

Incircle

Obviously, $\mathbf{n}$ is the midpoint of the incircle. Its radius r, the area $\triangle = \triangle_a + \triangle_b + \triangle_c$ and the **perimeter** $a + b + c$ of the triangle are related by

$$2\triangle = r(a + b + c).$$

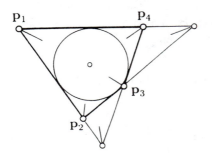

Quadrilateral about a Circle

Let $\mathbf{p}_1, \ldots, \mathbf{p}_4$ be the four vertices of a quadrilateral in the plane and let $s_{ik} = |\mathbf{p}_i - \mathbf{p}_k|$. If the quadrilateral is circumscribed about a circle, one has

$$s_{12} + s_{34} = s_{23} + s_{41}.$$

This follows from the fact that both tangents from a point to a circle have equal length.

The converse is also true if the quadrilateral is convex: Let C be the incircle touching the lines $\mathbf{p}_1\mathbf{p}_2$, $\mathbf{p}_2\mathbf{p}_3$ and $\mathbf{p}_3\mathbf{p}_4$, and let $\mathbf{p}_5$ be the point on $\mathbf{p}_3\mathbf{p}_4$ such that $\mathbf{p}_1\mathbf{p}_5$ touches C. Thus one has $s_{12} + s_{35} = s_{23} + s_{51}$ and, by assumption, also the identity above. This implies that $s_{45} + s_{14} = s_{15}$ and consequently $\mathbf{p}_4 = \mathbf{p}_5$.

Note that all six bisectors meet in one point, the center of the circle.

17.3 The Circumcircle

Another group of classical theorems is concerned with **perpendicular bisectors** of the sides of a triangle or quadrangle.

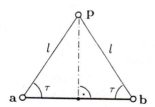

Midperpendicular

The **midperpendicular** of two points **a** and **b** consists of all points **p** such that

$$|\mathbf{p} - \mathbf{a}| = |\mathbf{p} - \mathbf{b}| \ .$$

The points **p** on the midperpendicular of **a** and **b** also satisfy

$$\measuredangle\mathbf{bap} = \measuredangle\mathbf{abp} = \tau \ .$$

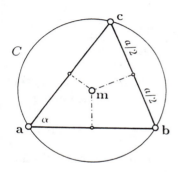

Circumcenter and Circumcircle

It follows that the three midperpendiculars of each triangle meet in a point **m** called the **circumcenter**. The circumcenter is the midpoint of the **circumcircle**. From the identities

$$2\triangle = bc\sin\alpha \ , \quad 2R\sin\alpha = a$$

etc., it follows that the radius R of the circumcircle is related to the area $\triangle$ of the triangle by

$$4R\triangle = abc \ .$$

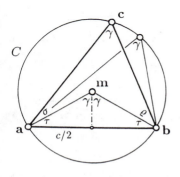

Angle of Circumference

For the angles ϱ, σ, τ shown in the figure one has

$$\alpha = \sigma + \tau, \quad \beta = \tau + \varrho, \quad \gamma = \varrho + \sigma \ .$$

Hence

$$\alpha + \varrho = \beta + \sigma = \gamma + \tau = 90° \ .$$

As a consequence, the **angle of circumference** γ does not change if **c** is moved along the circle, and it changes to $180° - \gamma$ if **c** crosses **a** or **b**.

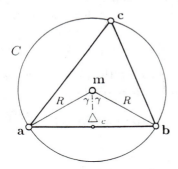

Barycentric Coordinates

From the equations $\triangle_c = R^2 \sin 2\gamma, \dots$, one gets that the **barycentric coordinates** of **m** with respect to **a**, **b**, **c** form the ratios

$$\sin 2\alpha : \sin 2\beta : \sin 2\gamma .$$

The angle **amb** $= 2\gamma$ is called the **central angle**.

Ptolemy's Theorem

Let $\mathbf{p}_1, \dots, \mathbf{p}_4$ be four points of a plane and let $s_{ik} = |\mathbf{p}_i - \mathbf{p}_k|$. Then one can show that $\mathbf{p}_1, \dots, \mathbf{p}_4$, in this order, lie on a circle if and only if

$$s_{31}\, s_{24} = s_{12}\, s_{34} + s_{23}\, s_{41} .$$

Note that all six midperpendiculars meet in one point, the center of the circle.

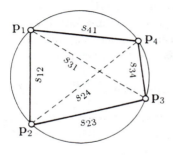

17.4 Power of a Point

In $\mathcal{E}^n$ a sphere is given in Cartesian coordinates by its normalized equation

$$S(\mathbf{x}) = \mathbf{x}^t\mathbf{x} - 2\mathbf{m}^t\mathbf{x} + c = 0 .$$

One can compute the sphere's its intersection with a straight line $\mathcal{L}$ given by $\mathbf{x} = \mathbf{p} + \mathbf{v}\lambda$, where $\mathbf{v}^t\mathbf{v} = 1$, to obtain a quadratic equation in λ,

$$\alpha\lambda^2 + \beta\lambda + \gamma = 0 ,$$

where $\alpha = \mathbf{v}^t\mathbf{v} = 1$ and $\gamma = S(\mathbf{p})$. Let λ_1 and λ_2 denote the equation's roots; then by Vieta's formula $\lambda_1\lambda_2 = \gamma$. The number $\gamma = S(\mathbf{p})$ is called the **power** of **p** with respect to S. The power has the following geometric interpretation.

Since $\mathbf{v}^t\mathbf{v} = 1$ the values λ form a **metric scale** on $\mathcal{L}$, i.e., one has $\lambda = |\mathbf{x} - \mathbf{p}|$. Thus $\gamma = \lambda_1\lambda_2$ is the product of the distances between each of the two intersection points and **p**. This product does not depend on **v**. In particular, if $\mathcal{L}$ is tangent to S, one has $\lambda_1 = \lambda_2$. Therefore $\sqrt{\gamma}$ is the distance between **p** and the point of contact of any tangent of S containing **p**. Note that **p** lies on S if $\gamma = 0$, and **p** lies inside of S if $\gamma < 0$.

Remark 1: By varying c one obtains a pencil of concentric spheres, where c presents the power of the origin.

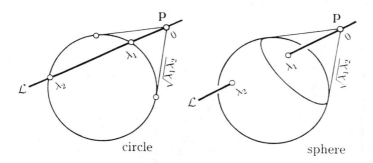

circle   sphere

Figure 17.1: Power of a point.

17.5 Radical Center

The powers of a point **x** with respect to two spheres S_1 and S_2 are equal if

$$\mathbf{x}^t\mathbf{x} - 2\mathbf{m}_1^t\mathbf{x} + c_1 = \mathbf{x}^t\mathbf{x} - 2\mathbf{m}_2^t\mathbf{x} + c_2 ,$$

i.e., if **x** lies in a hyperplane $\mathcal{P}_{12}$ defined by

$$2\left[\mathbf{m}_2 - \mathbf{m}_1\right]^t \mathbf{x} - (c_2 - c_1) = 0 .$$

This hyperplane is called the **radical plane** of both spheres or the **radical axis** if the spheres are circles. Note that this plane contains the real or non-real intersection of both spheres, where both powers vanish, and that it is perpendicular to the line spanned by the midpoints of both spheres.

For the three radical planes of any pair of the three spheres S_1, S_2, S_3 one has $\mathcal{P}_{12} + \mathcal{P}_{23} + \mathcal{P}_{31} = 0$, i.e., the three planes either meet in a line or are parallel.

It follows that in $\mathcal{E}^3$ the six radical planes corresponding to four spheres in general position meet at a point which is called the **radical center** of the four spheres.

Analogously, in $\mathcal{E}^2$ the three radical axes of three circles meet at a point, the radical center of the three circles, as illustrated in Figure 17.2. Note that this configuration can be viewed as the intersection of a plane with three spheres and their radical planes in $\mathcal{E}^3$.

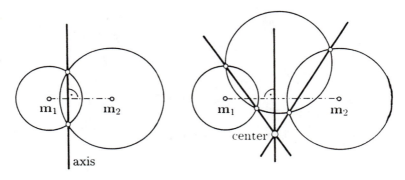

Figure 17.2: Radical axes and radical center of circles in the plane.

17.6 Orthogonal Spheres

The geometric meaning of the power $S(\mathbf{p})$ implies that each sphere around $\mathbf{p}$ with radius $\sqrt{S(\mathbf{p})}$ meets S orthogonally. Consequently, the midpoint of any sphere which meets two spheres S_1 and S_2 orthogonally lies in the radical plane of S_1 and S_2. There exists exactly one sphere orthogonal to four spheres given in general position in $\mathcal{E}^3$. Its midpoint $\mathbf{m}$ is their radical center; its radius is the root of the power of $\mathbf{m}$.

Analogous properties hold for orthogonal circles in the plane, as illustrated in Figure 17.3.

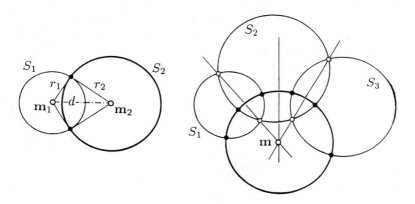

Figure 17.3: Orthogonal circles in the plane.

The radii r_1 and r_2 of two orthogonal spheres and the distance d of their midpoints are related by

$$d^2 = r_1^2 + r_2^2 \ .$$

Note that a sphere orthogonal to some given sphere S can have an imaginary radius. Its midpoint then lies inside of S.

17.7 Centers of Similitude

Any two spheres S_1 and S_2 with midpoints $\mathbf{m}_i$ and radii r_i are central affine. Namely, if $\mathbf{m}_1 + \mathbf{v}$ lies on S_1, then $\mathbf{m}_2 + \mathbf{v}\varrho$ lies on S_2, where

$\varrho = \pm r_2/r_1$. Solving the equation $\mathbf{m}_2 = \mathbf{c}(1 - \varrho) + \mathbf{m}_1\varrho$ one finds the centers of the dilatations are

$$\mathbf{c}_+ = \frac{\mathbf{m}_1 r_2 + \mathbf{m}_2 r_1}{r_2 + r_1} \qquad \text{and} \qquad \mathbf{c}_- = \frac{\mathbf{m}_1 r_2 - \mathbf{m}_2 r_1}{r_2 - r_1} \; .$$

This fact is illustrated in Figure 17.4 with circles.

Note that both centers are real even if the two spheres intersect, if S_2 lies in the interior of S_2, or if they are concentric. The center $\mathbf{c}_-$ is an ideal point only if $r_1 = r_2$. Note that $\mathbf{c}_+$ and $\mathbf{c}_-$ are the vertices of the tangent cones common to both spheres.

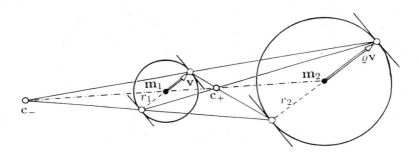

Figure 17.4: Both centers of similitude of two circles.

Three circles or spheres define three pairs of centers of similitude. These pairs form the opposite vertices of a complete quadrilateral, cf. Section 22.1. Every diagonal of the quadrilateral goes through two midpoints. In $\mathcal{E}^3$ the four sides of the quadrilateral are the intersections of the four symmetric pairs of planes tangent to all three spheres. These intersections lie in the plane of symmetry.

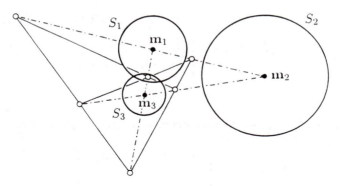

Figure 17.5: The six centers of similitude of three circles.

17.8 Notes and Problems

1 The orthocenter, the center of gravity, and the circumcenter of any triangle lie on one line (Euler).

2 The two bisectors of two lines in the plane are orthogonal.

3 In a plane, the midpoints of a family of circles touching two given circles lie on a conic section.

4 The family of spheres in $\mathcal{E}^3$ having the midpoints and radii of the circles of the family considered in Note 3 envelope a Dupin's cyclide, as presented in Section 19.7.

5 The distances $s_{ik} = |\mathbf{p}_i - \mathbf{p}_k|$ of five points $\mathbf{p}_1, \ldots, \mathbf{p}_5$ on a sphere satisfy

$$\det [s_{ik}] = 0 .$$

This generalizes Ptolemy's theorem to spheres.

6 The area $\triangle$ of a triangle spanned by $\mathbf{p}_1, \mathbf{p}_2, \mathbf{p}_3$ satisfies

$$-16\triangle^2 = \det \begin{bmatrix} 0 & s_{12}^2 & s_{13}^2 & 1 \\ s_{21}^2 & 0 & s_{23}^2 & 1 \\ s_{31}^2 & s_{32}^2 & 0 & 1 \\ 1 & 1 & 1 & 0 \end{bmatrix}.$$

7 The volume $\triangle$ of a tetrahedron spanned by $\mathbf{p}_1, \ldots, \mathbf{p}_4$ satisfies

$$288\triangle^2 = \det \begin{bmatrix} 0 & s_{12}^2 & \cdots & s_{14}^2 & 1 \\ s_{21}^2 & 0 & & \vdots & 1 \\ \vdots & & & s_{34}^2 & \vdots \\ s_{41}^2 & \cdots & s_{43}^2 & 0 & 1 \\ 1 & 1 & \cdots & 1 & 0 \end{bmatrix}.$$

8 Multiplying the radii of two spheres by a common factor does not influence the centers of similitude. In particular, one can regard the midpoints as spheres of radius 0.

18 Quadrics in Euclidean Space

Introducing distances and angles into an affine space also has several applications to quadrics. Normals and principal axes are Euclidean properties of quadrics which are invariant under Euclidean motions, but not under general affine maps. Moreover, one can distinguish quadrics in a Euclidean space by their semi-axes, i.e., by their shapes and sizes.

Literature: Berger, Blaschke, Greub

18.1 Normals

Consider a quadric Q given by the equation

$$Q(\mathbf{x}) = \mathbf{x}^t C \mathbf{x} + 2\mathbf{c}^t \mathbf{x} + c = 0 ,$$

where $\mathbf{x}$ denotes the Cartesian coordinates and $C^t = C$. In Section 16.4 it was shown that $\mathbf{u}$ represents the normal of the hyperplane $\mathbf{u}^t \mathbf{x} + u_0 = 0$, and one gets the following immediately:

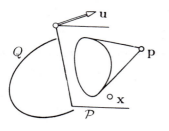

The **polar plane** $\mathcal{P}$ of a point $\mathbf{p}$ with respect to Q has the equation

$$[C\mathbf{p} + \mathbf{c}]^t \mathbf{x} + \mathbf{c}^t \mathbf{p} + c = 0 .$$

Hence, the normal of $\mathcal{P}$ is given by

$$\mathbf{u} = C\mathbf{p} + \mathbf{c} .$$

Note that $\mathcal{P}$ does not exist if $\mathbf{p}$ is a midpoint.

The **tangent plane** $\mathcal{T}$ of Q at a point $\mathbf{q} \in Q$ is the polar plane of $\mathbf{q}$ with respect to Q. Therefore

$$\mathbf{u} = C\mathbf{q} + \mathbf{c}$$

represents the normal of Q at $\mathbf{q}$. Recall that $\mathcal{T}$ is undefined if $\mathbf{q}$ is a singular point.

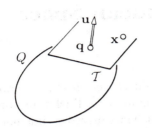

The **diametric plane** $\mathcal{V}$ of a direction $\mathbf{v}$ with respect to Q is defined by

$$[C\mathbf{v}]^t \mathbf{x} + \mathbf{v}^t \mathbf{c} = 0 \ .$$

Hence the normal $\mathbf{u}$ of $\mathcal{V}$ is given by

$$\mathbf{u} = C\mathbf{v} \ .$$

Recall that $\mathcal{V}$ does not exist if $\mathbf{v}$ is an axial direction.

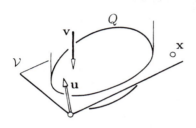

18.2 Principal Axes

If the diametric plane $\mathcal{V}$ with respect to $\mathbf{v}$ is perpendicular to $\mathbf{v}$, the quadric Q is symmetric to $\mathcal{V}$. Then $\mathbf{v}$ is called a **principal axis direction** or a **principal axis**.

In order to procure a principal axis $\mathbf{v}$ one can exploit the fact that $\mathbf{v}$ is parallel to the normal of its diametric plane, i.e., one has to solve the equation $C\mathbf{v} = \lambda\mathbf{v}$ or

$$[C - \lambda I]\mathbf{v} = \mathbf{o} \ .$$

This matrix equation represents a homogeneous linear system for the principal axes $\mathbf{v}$. It has non-trivial solutions only if

$$\det [C - \lambda I] = 0$$

which is a polynomial of proper degree n, called the **characteristic polynomial** of C. Its n roots λ_i are called the **eigenvalues** of C. For a given eigenvalue λ_i any non-trivial solution $\mathbf{v}_i$ of

$$[C - \lambda_i I]\,\mathbf{v}_i = \mathbf{o}$$

is called a **characteristic vector** or **eigenvector** associated with λ_i. The linear space spanned by all solutions $\mathbf{v}_i$ associated with λ_i is called a **characteristic space** or **eigenspace**. Eigenvalues coincide if they are multiple roots of the characteristic polynomial. Note that eigenvalues may be zero.

Remark 1: Because of their geometric definition, the principal axes of a quadric do not depend on the choice of the coordinate system.

18.3 Real and Symmetric Matrices

If the matrix C is real and symmetric, eigenvalues and eigenvectors have particularly nice properties:

The eigenvectors corresponding to different eigenvalues are orthogonal.
The eigenvalues are real.
The eigenspace associated with an r-fold eigenvalue has dimension r.

In order to prove these facts consider first two arbitrary eigenvectors $\mathbf{v}_1$ and $\mathbf{v}_2$ associated with two different eigenvalues λ_1 and λ_2, respectively. One has

$$C\mathbf{v}_1 = \mathbf{v}_1\lambda_1 \quad \text{and} \quad C\mathbf{v}_2 = \mathbf{v}_2\lambda_2 \ ,$$

and since $\mathbf{v}_2^t C\mathbf{v}_1 = \mathbf{v}_1^t C\mathbf{v}_2$ one also has $\mathbf{v}_2^t \mathbf{v}_1\lambda_1 = \mathbf{v}_1^t \mathbf{v}_2\lambda_2$, i.e.,

$$\mathbf{v}_2^t \mathbf{v}_1(\lambda_2 - \lambda_1) = 0 \ .$$

If $\lambda_1 \neq \lambda_2$, one has $\mathbf{v}_2^t \mathbf{v}_1 = 0$ which means that $\mathbf{v}_1$ and $\mathbf{v}_2$ are orthogonal.

Secondly, assume that $\lambda_1 = \alpha + i\beta$, where $\beta \neq 0$ is a non-real eigenvalue, and $\mathbf{v}_1 = \mathbf{a} + i\mathbf{b}$ is the associated eigenvector. Then $\lambda_2 = \alpha - i\beta$ also is an eigenvalue, and $\mathbf{v}_2 = \mathbf{a} - i\mathbf{b}$ is the associated eigenvector. Since

$$\mathbf{v}_1^t \mathbf{v}_2 = \mathbf{a}^t\mathbf{a} + \mathbf{b}^t\mathbf{b} > 0 \ ,$$

one has $\lambda_1 = \lambda_2$. Thus $\beta = 0$ and λ_1 is real.

Finally, assume that λ_1 is an r-fold eigenvalue and $\mathbf{v}_1$ is an associated eigenvector. Then one can introduce a new orthonormal basis $\mathbf{b}_1, \ldots, \mathbf{b}_n$ such that $\mathbf{b}_1$ is parallel to $\mathbf{v}_1$. Thus, one gets

$$\overline{C} = B^t C B = \begin{bmatrix} \lambda_1 & 0 & \cdots & 0 \\ 0 & & & \\ \vdots & & C^* & \\ 0 & & & \end{bmatrix}, \quad B = [\mathbf{b}_1 \ldots \mathbf{b}_n] .$$

Because of Remark 1, C has the eigenvalues C^* and λ_1. Hence, λ_1 is an $(r-1)$-fold eigenvalue of C^*. Since C^* is symmetric and real, one can perform this reduction step r times thereby obtaining r (linearly independent) orthonormal eigenvectors associated with λ_1.

18.4 Principal Axis Transformation

The matrix C of a **quadratic form** $\mathbf{v}^t C \mathbf{v} = 0$ in $\mathbf{E}^n$ has n orthogonal eigenvectors, as is shown in the previous section. Thus one can construct an orthonormal basis $\mathbf{b}_1, \ldots, \mathbf{b}_n$ of $\mathbf{E}^n$ consisting of the eigenvectors of C such that the determinant of $B = [\mathbf{b}_1 \ldots \mathbf{b}_n]$ equals $+1$. The associated Euclidean transformation

$$\mathbf{v} = B\mathbf{w}$$

is called a **principal axis transformation**. The equation of the transformed quadratic form is $\mathbf{w}^t \overline{C} \mathbf{w}$ where $\overline{C} = B^t C B$. Let Λ denote the diagonal matrix of the eigenvalues λ_i corresponding to $\mathbf{b}_1, \ldots, \mathbf{b}_n$, i.e.,

$$\Lambda = \begin{bmatrix} \lambda_1 & & \\ & \ddots & \\ & & \lambda_n \end{bmatrix} \quad \text{and} \quad CB = B\Lambda .$$

Consequently one has

$$\overline{C} = B^t C B = B^t B \Lambda = \Lambda .$$

In words: The principal axis transformation transforms the quadratic form $\mathbf{v}^t C \mathbf{v}$ into the diagonal form $\mathbf{w}^t \Lambda \mathbf{w}$. Note that the principal axis transformation is not unique.

Remark 2: One can compute the eigenspace associated with some eigenvalue by means of the Gauss-Jordan algorithm. Subsequently one can construct an orthonormal basis of the eigenspace by, e.g., the Gram-Schmidt orthonormalization.

Remark 3: In view of Section 13.2 the eigenvectors associated with a zero eigenvalue of C are called **proper axis directions** of the **quadratic form $\mathbf{v}^t C \mathbf{v} = 0$.**

18.5   Normal Forms of Euclidean Quadrics

Let $\mathbf{x}^t C \mathbf{x} + 2\mathbf{c}^t \mathbf{x} + c = 0$ be the equation of some arbitrary quadric Q in $\mathcal{E}^n$. Zero entries of C correspond to conjugate pairs of coordinate directions. In the sequel, it is shown that one can construct a Cartesian system such that all coordinate directions are conjugate with respect to Q and that the equation of Q takes on one of the following three normal forms

$$\mu_1 x_1^2 + \cdots + \mu_r x_r^2 = \begin{cases} 1, & 1 \le r \le n \\ 0, & 1 \le r \le n \\ 2x_n, & 1 \le r < n \end{cases}.$$

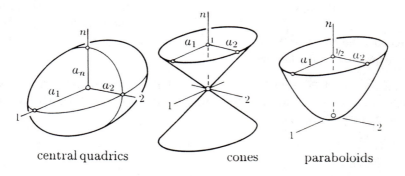

central quadrics   cones   paraboloids

Figure 18.1: Euclidean normal forms.

Figure 18.1 gives an illustration. Recall from Section 14.4 that the first normal form represents a central quadric, where $\mathcal{M} \neq \emptyset$ but $\mathcal{S} = \emptyset$; the second normal form represents a cone, where $\mathcal{M} = \mathcal{S} \neq \emptyset$; and the third normal form represents a paraboloid, where $\mathcal{M} = \emptyset$.

The underlying Cartesian system is constructed in the following way:

First, after a principal axis transformation $\mathbf{x} = B\mathbf{y}$, which is discussed in the previous section, one obtains the equation

$$\overline{Q}(\mathbf{y}) = \mathbf{y}^t \Lambda \mathbf{y} + 2\overline{\mathbf{c}}\mathbf{y} + c = 0,$$

where $\overline{\mathbf{c}} = B^t\mathbf{c}$.

Second, one can rid the equation of the linear term by the translation $\mathbf{y} = \mathbf{b} + \mathbf{z}$, where

$$\Lambda \mathbf{b} + \overline{\mathbf{c}} = \mathbf{o} \ ,$$

provided that there is a solution $\mathbf{b}$. Subsequently one can multiply the equation of the quadric by a suitable factor in order to procure the normal form as a function of $\mathbf{z}$.

However, if $\Lambda \mathbf{b} + \overline{\mathbf{c}} = \mathbf{o}$ has no solution, the quadric is a paraboloid. Then one constructs the orthonormal basis $\mathbf{b}_1, \ldots, \mathbf{b}_n$ for the principal axis transformation such that $\mathbf{b}_{r+1}, \ldots, \mathbf{b}_n$ span the eigenspace $\mathbf{V}_0$ associated with the zero eigenvalue, where $\mathbf{b}_{r+1}, \ldots, \mathbf{b}_{n-1}$ are orthogonal to $\mathbf{c}$. This is possible since $\mathbf{c}$ is not an element of $\mathbf{V}_0$.

After performing the axis transformation above, one solves the first $n-1$ equations of the linear system $\Lambda \mathbf{b} + \overline{\mathbf{c}} = \mathbf{o}$ and chooses the last coordinate b_n of $\mathbf{b}$ such that $\mathbf{b}^t \Lambda \mathbf{b} + 2\overline{\mathbf{c}}^t \mathbf{b} + c = 0$. Then, the translation $\mathbf{y} = \mathbf{b} + \mathbf{z}$ changes the equation $\overline{Q}(\mathbf{y}) = 0$ into the normal form depending on $\mathbf{z}$ and multiplied by some factor.

Remark 4: The coefficients μ_i above represent the non-vanishing eigenvalues of the normal forms. On setting $\mu_i = \pm 1/a_i^2$ the three normal forms can also be written as

$$\pm\frac{x_1^2}{a_1^2} \pm \cdots \pm \frac{x_r^2}{a_r^2} = \begin{cases} 1 & 1 \leq r \leq n \\ 0 & 1 \leq r \leq n \\ 2x_n & 1 \leq r < n \ . \end{cases}$$

The positive numbers a_i are called the **semi-axes** of the respective quadric. An axis is called **real** if $\mu_i > 0$. Otherwise it is called a **non-real** or **imaginary axis**.

Example 1: Consider the quadric in $\mathcal{E}^3$ given by

$$C = \begin{bmatrix} 1 & 0 & 0 \\ 0 & 0 & 0 \\ 0 & 0 & 0 \end{bmatrix} , \quad \mathbf{c} = \begin{bmatrix} -1 \\ 0 \\ 0 \end{bmatrix} , \quad c = 0 .$$

The three coordinate axes are already parallel to the eigenvectors of C, i.e., $\Lambda = C$. Obviously, $\mathbf{b} = -\mathbf{c}$ is a solution of $\Lambda\mathbf{b} + \mathbf{c} = \mathbf{o}$. Hence, $\mathbf{x} = \mathbf{b} + \mathbf{y}$ transforms Q into $y_1^2 = 1$. This example is illustrated in the left part of Figure 18.2.

Example 2: Consider the quadric in $\mathcal{E}^3$ given by

$$C = \begin{bmatrix} 1 & 0 & 0 \\ 0 & 0 & 0 \\ 0 & 0 & 0 \end{bmatrix} , \quad \mathbf{c} = \begin{bmatrix} -1 \\ -1/4 \\ -1/4 \end{bmatrix} , \quad c = 0 .$$

The three coordinate axes are already parallel to the eigenvectors of C, i.e., $\Lambda = C$, but $\Lambda\mathbf{m} + \mathbf{c} \neq \mathbf{o}$ for all $\mathbf{m}$. However, the basis vectors associated with the zero eigenvalue, the 2- and 3-direction, are not orthogonal to $\mathbf{c}$. Therefore these vectors are replaced by

$$\mathbf{b}_2 = [0 \ \varrho \ -\varrho]^t , \quad \mathbf{b}_3 = [0 \ \varrho \ \varrho]^t \quad \text{where } \varrho = \sqrt{2}/2 .$$

Then the quadric with respect to this new system is given by

$$\Lambda = \begin{bmatrix} 1 & 0 & 0 \\ 0 & 0 & 0 \\ 0 & 0 & 0 \end{bmatrix} , \quad \bar{\mathbf{c}} = \begin{bmatrix} -1 \\ 0 \\ -\varrho/2 \end{bmatrix} , \quad \bar{c} = 0 .$$

Obviously, $\mathbf{b} = [1 \ 0 \ \delta]^t$ solves the first two equations of $\Lambda\mathbf{b} + \bar{\mathbf{c}} = \mathbf{o}$, and the transformation $\mathbf{y} = \mathbf{b} + \mathbf{z}$ changes the equation of Q into $z_1^2 - \varrho z_3 - (1 + \varrho\delta) = 0$ which is the normal form for $\delta = -1/\varrho$. This example is illustrated in the right part of Figure 18.2.

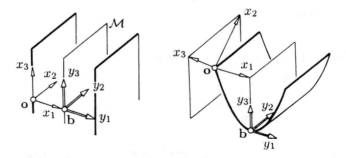

Figure 18.2: Normal form, examples.

18.6 Notes and Problems

1 For $n = 2$ the angles α between the principal axes and the x-axis satisfy

$$\tan 2\alpha = \frac{2c_{1,2}}{c_{1,1} - c_{2,2}} \ .$$

This follows from the coordinate transformation $\mathbf{x} = B\mathbf{y}$,

$$B = \begin{bmatrix} \cos\alpha & -\sin\alpha \\ \sin\alpha & \cos\alpha \end{bmatrix} \ ,$$

where $\overline{C} = B^t C B$ becomes diagonal.

2 Homothetic quadrics in a Euclidean space have the same principal axes directions.

3 In $\mathcal{E}^n$ homothetic quadrics are represented by their normal form

$$\mu_1^2 x_1^2 + \cdots + \mu_r x_r^2 = \begin{cases} c & , \quad 1 \le r \le n \\ 2x_n + c & , \quad 1 \le r < n \ . \end{cases}$$

4 Parallel intersections of a quadric have parallel principal axes directions.

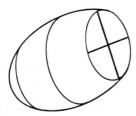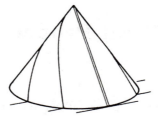

Figure 18.3: Parallel intersections.

5 Since the affine normal form can easily be derived from the Euclidean one, the Euclidean method can also be used in affine spaces to construct the affine normal form of a quadric.

6 Note 4 also verifies the existence of the affine normal forms in Section 14.4.

7 A matrix C also represents a linear map φ. This means that the eigenvectors of C are the vectors mapped onto certain multiples of themselves under φ. This interpretation characterizes eigenvectors and eigenvalues without reference to a basis. Hence, each matrix representing φ has the same eigenvalues and the same eigenvectors, only represented in different systems.

8 A point of a quadric where the normal is parallel to a principal axis direction is called a **vertex**.

19 Focal Properties

Euclidean measures allow, for example, the notion of principal axes. Other examples of Euclidean concepts are the properties of focal points of quadrics. Focal points are crucial in optical systems and for constructing quadrics and Dupin's cyclides in geometric modeling.

Literature: Berger, Blaschke, *Maxwell*, Straßer-Seidel

19.1 The Ellipse

An ellipse given by the equation

$$\frac{x^2}{a^2} + \frac{y^2}{b^2} = 1 \ , \qquad a > b \ ,$$

is represented in parametric form by

$$x = a \cos \varphi \ , \qquad y = b \sin \varphi \ .$$

Let $e^2 = a^2 - b^2$. The points $\mathbf{f}_\pm = [\pm e \quad 0]^t$ are called the **foci** of the ellipse. These points are characteristic Euclidean features of the ellipse, as demonstrated by the following three properties:

Let l_+ and l_- be the distances between a point $\mathbf{x}$ on the ellipse and the foci $\mathbf{f}_+$ and $\mathbf{f}_-$, respectively. Then some elementary calculations give

$$l_+ = a - e\cos\varphi = a - \frac{e}{a}x \ ,$$

$$l_- = a + e\cos\varphi = a + \frac{e}{a}x \ .$$

Therefore the sum of both distances,

$$l = l_+ + l_- = 2a \ ,$$

does not depend on $\mathbf{x}$ which gives rise to the well-known **string construction** of the ellipse illustrated in Figure 19.1.

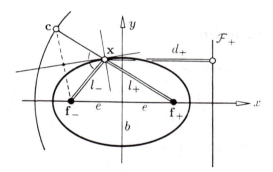

Figure 19.1: Focal properties of an ellipse.

The tangent and normal of the ellipse at φ have the equations

$$\frac{\cos\varphi}{a}x + \frac{\sin\varphi}{b}y = 1 \quad \text{and} \quad \frac{a}{\cos\varphi}x - \frac{b}{\sin\varphi}y = e^2 \ ,$$

respectively. Let $\mathbf{c} = [\mathbf{x} \cdot 2a - \mathbf{f}_+ \cdot l_-]/l_+$ be the point on the ray from $\mathbf{x}$ through $\mathbf{f}_+$ at distance l from $\mathbf{f}_+$. One can check that the midpoint between $\mathbf{c}$ and $\mathbf{f}_-$ lies on the tangent at $\mathbf{x}$. This means that the tangent bisects the angle between both parts of the string, as illustrated in Figure 19.1.

The polar lines of the foci are called **focal lines**. They have the equations $x = \pm a^2/e$. Let d_+ denote the distance between $\mathbf{x}$ and the focal line $x = a^2/e$ which is denoted by $\mathcal{F}_+$. Then one gets

$$d_+ = \frac{a^2}{e} - x = \frac{a}{e}\left(a - \frac{e}{a}x\right) = \frac{a}{e}l_+$$

and the so-called **numerical eccentricity** of the ellipse

$$\varepsilon = \frac{l_+}{d_+} = \frac{e}{a} < 1 \ .$$

The numerical eccentricity is constant for all points $\mathbf{x}$ of the ellipse, and it characterizes the ellipse as does the string construction above. This is also illustrated in Figure 19.1.

19.2 The Hyperbola

A hyperbola given by the equation

$$\frac{x^2}{a^2} - \frac{y^2}{b^2} = 1 \ , \qquad a > b \ ,$$

is represented in parametric form by

$$x = a/\cos\psi \ , \qquad y = b\tan\psi \ .$$

Here, let $e^2 = a^2 + b^2$. The points $\mathbf{f}_\pm = [\pm e \ 0]^t$ are called the **foci** of the hyperbola. Let l_+ and l_- be the distances between a point $\mathbf{x}$ on the hyperbola and the foci $\mathbf{f}_+$ and $\mathbf{f}_-$, respectively. Then some calculations show that

$$l_+ = e/\cos\psi - a = \frac{e}{a}x - a \ ,$$

$$l_- = e/\cos\psi + a = \frac{e}{a}x + a \ .$$

It follows that the difference $l_- - l_+ = 2a$ does not depend on $\mathbf{x}$, which verifies the well-known **string construction** of a hyperbola as illustrated in Figure 19.2.

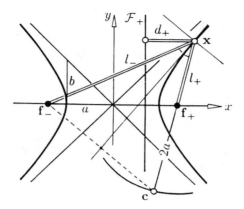

Figure 19.2: Focal properties of a hyperbola.

The tangent and the normal of the hyperbola at ψ have the equations

$$\frac{1}{a\cos\psi}\,x - \frac{\tan\psi}{b}\,y = 1 \quad \text{and} \quad a\cos\psi\cdot x + \frac{b}{\tan\psi}\cdot y = e^2 \;,$$

respectively. Let $\mathbf{c} = [\mathbf{f}_+ \cdot l_- - \mathbf{x}\cdot 2a]/l_+$ be the point on the ray from $\mathbf{x}$ through $\mathbf{f}_+$ at distance l_- from $\mathbf{f}_+$. Using the same argument as above one obtains that the tangent at $\mathbf{x}$ bisects the angle between both parts of the string.

The polar lines of the foci are called **focal lines**. They have the equations $x = \pm a^2/e$. Let d_+ denote the distance between $\mathbf{x}$ and the focal line $x = a^2/e$. Then, one gets

$$d_+ = x - \frac{a^2}{e} = \frac{a}{e}\left(\frac{e}{a}x - a\right) = \frac{a}{e}l_+$$

and the so-called **numerical eccentricity** of the hyperbola

$$\varepsilon = \frac{l_+}{d_+} = \frac{e}{a} > 1 \;.$$

The numerical eccentricity is constant for all points $\mathbf{x}$ of the hyperbola, and it characterizes the hyperbola as does the string construction above.

19.3 The Parabola

A parabola given by the equation

$$2px = y^2 , \qquad p > 0 ,$$

is also represented in parametric form by

$$y = 2pt , \qquad x = 2pt^2 .$$

The point $\mathbf{f} = [p/2 \;\; 0]^t$ is called the **focus** of the parabola. Some calculations show that the distance l between a point $\mathbf{x}$ on the parabola and $\mathbf{f}$ satisfies

$$l = x + \frac{p}{2} .$$

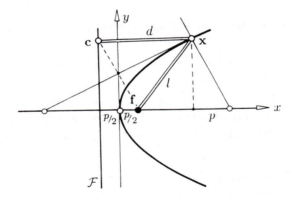

Figure 19.3: Focal properties of a parabola.

The polar line $\mathcal{F}$ of the focus is called **focal line**. It has the equation $x = -p/2$. Let d denote the distance between $\mathbf{x}$ and the focal line; then one gets

$$d = x + \frac{p}{2} = l .$$

This relationship holds for any point $\mathbf{x}$ of the parabola, and it verifies the well-known **string construction** of the parabola. It is illustrated in Figure 19.3.

The tangent and the normal of the parabola at t intersect the x-axis at the points $[-x \quad 0]^t$ and $[x+p \quad 0]^t$, each of which is at a distance of l from $\mathbf{f}$. The foot of the perpendicular from $\mathbf{x}$ onto the focal line is $\mathbf{c} = [-p/2 \quad y]^t$. One can check again that the midpoint of $\mathbf{c}$ and $\mathbf{f}$ lies on the tangent at $\mathbf{x}$. This means that the tangent bisects the angle between both parts of the string.

The **numerical eccentricity**, $\varepsilon = l/d$, of a parabola equals one.

19.4 Confocal Conic Sections

Consider the family of conic sections in the plane given by

$$\frac{x^2}{a^2 - \lambda} + \frac{y^2}{b^2 - \lambda} = 1 \ ,$$

where λ varies and $a^2 - b^2 = e^2$. This equation represents for

$$\lambda < b^2 \quad \text{an ellipse and for}$$
$$b^2 < \lambda < a^2 \quad \text{a hyperbola.}$$

All these curves are **confocal**, i.e., they have the same foci $[\pm e \quad 0]^t$. The foci themselves can be regarded as degenerate members of the family, where $\lambda = b^2$. Figure 19.4 gives an illustration. The conic sections of the family have the parametric representation

$$x = e \frac{\cos \varphi}{\cos \psi} \ , \qquad y = e \sin \varphi \tan \psi \ ,$$

where the isolines $\varphi = \textit{fixed}$ and $\psi = \textit{fixed}$ describe the confocal hyperbolas $\lambda = b^2 + e^2 \sin^2 \varphi$ and ellipses $\lambda = b^2 - e^2 \tan^2 \psi$, respectively. From the string constructions it follows immediately that the isolines meet orthogonally.

Similarly, the parabolas of the family

$$\frac{2x}{\lambda} + \frac{y^2}{\lambda^2} = 1 \ ,$$

where λ varies, are confocal. The origin is the common focus and the x-axis the common axis. The parabolas of the family have the parametric representation

$$2x = \lambda_1 + \lambda_2 , \qquad y^2 = -\lambda_1 \lambda_2 ,$$

where the isolines $\lambda_1 = \textit{fixed} < 0$ and $\lambda_2 = \textit{fixed} > 0$ represent confocal parabolas which meet orthogonally. Besides the position in the plane there is only one family of confocal parabolas, which is shown in the right part of Figure 19.4.

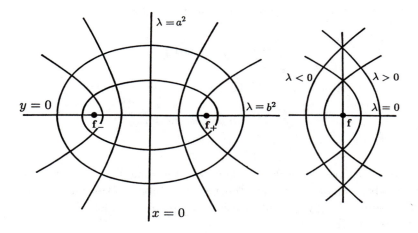

Figure 19.4: Confocal conic sections.

Remark 1: Trigonometric functions of φ are known to be rational quadratic functions of

$$t = \tan \frac{\varphi}{2} = \frac{u}{1-u} ,$$

where $u = 0, \frac{1}{2}, 1$ corresponds to $\varphi = 0°, 90°, 180°$ and $t = 0, 1, \infty$. Using the quadratic Bernstein polynomials

$$B_0(u) = (1-u)^2 , \qquad B_1(u) = 2(1-u)u , \qquad B_2(u) = u^2$$

one has, for example,

$$\sin\varphi = \frac{2t}{1+t^2} = \frac{B_1}{B_0+B_2} \,, \qquad \cos\varphi = \frac{1-t^2}{1+t^2} = \frac{B_0-B_2}{B_0+B_2} \,.$$

With the abbreviations $B_i(u) = U_i$ and $B_i(v) = V_i$ this transformation can be applied to the parametric representation of confocal ellipses and hyperbolas which in homogeneous coordinates results in

$$xw = e(U_0 - U_2)(V_0 + V_2) \,,$$
$$yw = eU_1V_1 \,,$$
$$w = (U_0 + U_2)(V_0 - V_2) \,.$$

This representation is called a rational **Bernstein-Bézier representation**. It is evaluated most effectively by the algorithm of de Casteljau in Section 12.3.

19.5 Focal Conics

A cone which circumscribes a sphere is called a **right cone**. Using the string constructions of ellipses and hyperbolas it can be shown that the vertices **h** of all real right cones which contain an ellipse E given by

$$\frac{x^2}{a^2} + \frac{y^2}{b^2} = 1 \,, \qquad z = 0 \,, \qquad a > b \,,$$

lie on the hyperbola H defined by

$$\frac{x^2}{e^2} - \frac{z^2}{b^2} = 1 \,, \qquad y = 0 \,, \qquad e^2 = a^2 - b^2 \,.$$

Similarly, the vertices of all real right cones containing H lie on E.

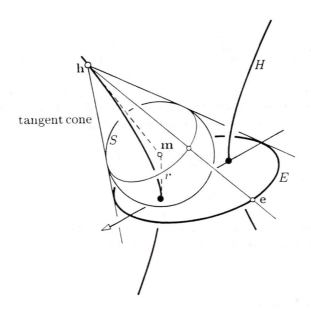

Figure 19.5: Right cone to an ellipse.

Moreover, it can be shown that every cone circumscribing a quadric Q of the family

$$\frac{x^2}{a^2 - \lambda} + \frac{y^2}{b^2 - \lambda} + \frac{z^2}{-\lambda} = 1$$

is a right cone if its vertex is on E or H. E and H are called the **focal conics** of Q, and the family of quadrics above is called **confocal**. In particular, one has for

$$\begin{aligned} \lambda &< 0 & &\text{ellipsoids,} \\ 0 < \lambda &< b^2 & &\text{hyperboloids of one sheet,} \\ b^2 < \lambda &< a^2 & &\text{hyperboloids of two sheets,} \end{aligned}$$

as illustrated in Figure 19.6. For $\lambda > a^2$ the members of the family have no real points. The focal conics E and H can be viewed as degenerate members of this family, where $\lambda = 0$ and $\lambda = -b^2$, respectively.

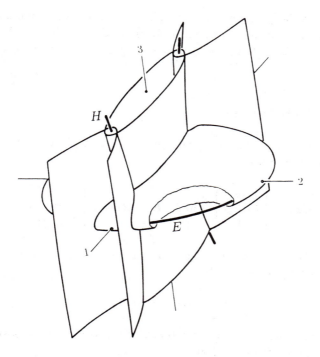

Figure 19.6: Confocal central quadrics.

For any given point $[x \quad y \quad z]^t$ there are three members of the family containing this point. Furthermore, any two (or three) members $Q_1(\mathbf{x}) = 0, Q_2(\mathbf{x}) = 0$ of the family intersect each other orthogonally since at a common point the scalar product of the normals of Q_1 and Q_2 equals a multiple of $Q_1 - Q_2 = 0$.

Remark 2: As a consequence of the right cone property above, E and H meet each quadric of the confocal family in its **umbilical points** (see Sections 15.6 and 32.1). In particular, if the interior of E is regarded as a very flat ellipsoid, the foci of E can be viewed as the umbilical points of this flat ellipsoid, and analogously for H. Note that the foci of E and H are the vertices of H and E, respectively.

Remark 3: Similarly, the vertices of right cones which are circumscribed about a paraboloid lie on two so-called **focal parabolas**. The parabolas lie in perpendicular planes, and the focus of each parabola is the vertex of the other one.

19.6 Focal Distances

The points $\mathbf{e}$ and $\mathbf{h}$ of the two focal conic sections E and H above have the parameter representations

$$\mathbf{e} = \mathbf{e}(\varphi) = \begin{bmatrix} a\cos\varphi \\ b\sin\varphi \\ 0 \end{bmatrix} \quad \text{and} \quad \mathbf{h} = \mathbf{h}(\psi) = \begin{bmatrix} e/\cos\psi \\ 0 \\ b\tan\psi \end{bmatrix} ,$$

respectively. By some standard algebraic manipulations one can compute the distance d between $\mathbf{e}$ and $\mathbf{h}$,

$$d = a/\cos\psi - e\cos\varphi .$$

Note that d has a sign. One has $d > 0$ or $d < 0$ if $\mathbf{h}$ lies on the positive or negative branch of H, respectively.

Example 1: Let $\mathbf{h}_0 = \mathbf{h}(\psi_0)$ and $\mathbf{h} = \mathbf{h}(\psi)$ be two points on different branches of H, as illustrated in Figure 19.7. Their positive distances to some point $\mathbf{e}$ on E are

$$d_0 = e\cos\varphi - a/\cos\psi_0 ,$$
$$d = a/\cos\psi - e\cos\varphi .$$

Obviously, the sum $d_0 + d_1$ is independent of $\mathbf{e}$ on E. This generalizes the string construction in Section 19.1.

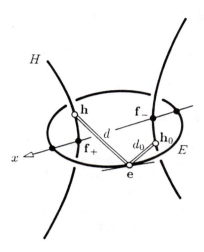

Figure 19.7: Focal conics and generalized string construction of E.

Remark 4: From the right cone property in Section 19.6 or just by mechanical reasoning it follows that the angles between the tangent at e and either part of the string from h_0 through e to h_1 are equal.

19.7 Dupin's Cyclide

A line meeting both focal conics is called a **focal ray**. Let the points e, h_0, and h be the same as above, and let z be an additional point on the focal ray through e and h, with the distances

$$d_e = \kappa - e\cos\varphi \quad \text{and} \quad d_h = a/\cos\varphi - \kappa$$

between z and e and between z and h, respectively, such that $d_e + d_h = d$. If e and h vary with φ and ψ, but κ remains fixed, the point z traces out a surface, called a **Dupin's cyclide**. It is called confocal to E and H and given by the affine combination

$$\mathbf{z} = \mathbf{z}(\varphi, \psi) = \frac{1}{d}[\mathbf{e} \cdot d_h + \mathbf{h} \cdot d_e] \ .$$

Consider d_0 of Example 1 where $\psi_0 = 0$ for simplicity. The sum

$$l = d_0 + d_e = a - \kappa$$

does not depend on $\mathbf{e}$ or on $\mathbf{h}$.

This has a nice and simple geometric interpretation due to Maxwell (1868). Let a sufficiently long string be fastened at one end to the focus of an ellipse (or a hyperbola or parabola) C, and let the string slide smoothly over this conic section while keeping it always tight, then the unfastened end will sweep out a **Dupin cyclide** Z confocal to C. In particular, if C is a parabola, the cyclide is called a **parabolic cyclide**.

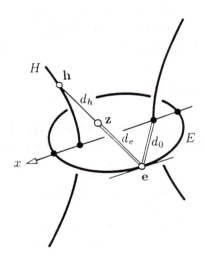

Figure 19.8: String construction of Dupin's cyclide.

For mechanical reasons the focal ray **eh** is normal to Z at **z**. Moreover, the cyclide is enveloped by a family of spheres around the points **e** of E with radii d_e and a second family of spheres around the points **h** of H with radii d_h.

On each focal ray the values of κ form a metric scale. As a consequence one has:

Confocal Dupin's cyclides are offsets from each other.

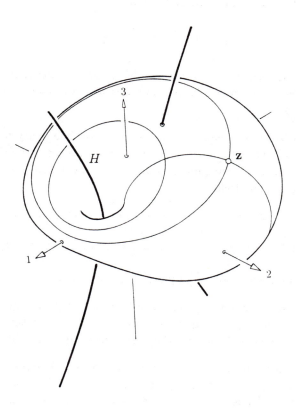

Figure 19.9: Dupin's cyclide, example.

Example 2: If C is a circle, the string is fastened at the center of the circle. Then the free end of the tight string sweeps out a **torus**.

Remark 5: Introducing $t = \tan \varphi/2$ and $s = \tan \psi/2$ as in Remark 1, the parametric representation of a Dupin's cyclide becomes rational quadratic in s and t which can easily be transformed into a Bernstein-Bézier representation.

19.8 Notes and Problems

1 If the focus $\mathbf{f}_+$ of a conic lies at the origin, the conic has the following representation in polar coordinates r, ϑ:

$$r(\vartheta) = \frac{p}{1 + \varepsilon \cos \vartheta} \ ,$$

where the numerical eccentricity $\varepsilon = e/a$, and the **parameter** $p = b^2/a$ for an ellipse or a hyperbola; for a parabola $\varepsilon = 1$, and **p** is defined in Section 19.3.

2 A confocal family of quadrics also contains two imaginary focal conics. For $\lambda = -a^2$ one gets an imaginary conic in the plane $x = 0$, and for $\lambda = \infty$ one gets an imaginary conic in the ideal plane, the so-called **absolute spherical circle**.

3 Let δ denote the angle between string and tangent, then one has

$$\tan \delta = b/e \cdot \sin \varphi \ , \quad \tan \delta = b/e \cdot \tan \psi \ , \quad \tan \delta = 1/2t$$

for the ellipse, hyperbola, and parabola, respectively.

4 A family of confocal ellipses and hyperbolas also contains two imaginary pairs of foci. For $\lambda = a^2$ one gets an imaginary pair on the y-axis, and for $\lambda = \infty$ one gets the so-called **cyclic points**, which are ideal.

5 Light emanating from one focus $\mathbf{f}$ of an ellipse C is reflected by C to the other focus. In case of a hyperbola the reflected light seems to come from the other focus. In case of a parabola the reflected light rays are parallel.

6 In principal axes coordinates **confocal cones** are defined by

$$\frac{x^2}{a^2 - \lambda} + \frac{y^2}{b^2 - \lambda} + \frac{z^2}{-\lambda^2} = 0 \ .$$

7 The tangent cones from any point to a family of confocal quadrics form a family of confocal cones.

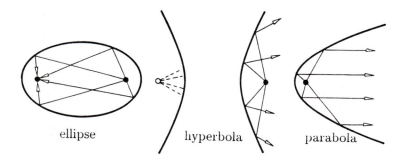

ellipse   hyperbola   parabola

Figure 19.10: Reflections of focal rays.

8 A family of confocal cones contains one pair of real focal rays and two imaginary pairs.

9 A family of confocal cones intersects a concentric sphere in a family of so-called **spherical conic sections**.

10 Spherical conic sections can also be constructed with a tight string sliding on a sphere.

11 Light emanating from the focal ellipse E in direction of a focal ray to a confocal ellipsoid Q is reflected in the direction of a focal ray back to E.

12 A system of confocal cyclides can also be regarded as a system of wave surfaces in an isotropic medium corresponding to the focal rays from a fixed point on H (or E) reflected by E (or H).

13 In 1868 Maxwell deduced a string construction of space quadrics from the properties mentioned in Notes 11 and 12.

PART FIVE

Some Projective Geometry

Parallelism is a characterizing concept of affine spaces. However, the fact that coplanar lines may or may not intersect gives rise to many special cases in affine geometry. Therefore, the invention of points at infinity by the French architect Girard Desargues (1591–1661) has lead to a unified approach and deeper insight into geometric structures. In the words of the English mathematician Arthur Cayley (1821–1895): "Projective geometry is all geometry."

In projective geometry, the points at infinity are not distinguished from the other points. This is the reason why projective geometry exhibits much more symmetry than affine geometry. With the introduction of homogeneous coordinates by Plücker (1801–1868) it became possible to study projective geometry analytically. Barycentric coordinates are a specialization of homogeneous coordinates whose symmetry became apparent in the discussion of affine geometry in Part Three.

20 The Projective Space

Homogeneous coordinates, which were introduced in Chapter 6, are the key to projective spaces. These coordinates can be viewed as generalized barycentric coordinates by suspending the condition that they form a partition of one.

Literature: Baker, Berger, Samuel, Wylie

20.1 Homogeneous Coordinates

Projective spaces can be viewed as extensions of affine spaces, as in Chapter 6. Then $\mathbf{y}^t = [y_0 \quad \mathbf{y}^t]$ represents the point

$$\mathbf{x} = \frac{\mathbf{y}}{y_0}$$

of some n-dimensional affine space $\mathcal{A}$ if $y_0 \neq 0$, and it represents the direction of some line

$$\mathbf{x} = \mathbf{b} + \mathbf{y}\lambda$$

in $\mathcal{A}$ if $y_0 = 0$.

Following this idea the elements of an n-dimensional projective space $\mathcal{P}$ can be defined as the one-dimensional subspaces of some $n+1$ dimensional vector space. Hence, the elements of $\mathcal{P}$, called **points**, are represented by homogeneous coordinate columns

$$
y = \begin{bmatrix} y_0 \\ \vdots \\ y_n \end{bmatrix} ,
$$

where y and $y\varrho$ represent the same point of $\mathcal{P}$ for all $\varrho \neq 0$. Note that the zero column $y = o = [0 \ldots 0]^t$ does not represent any point of $\mathcal{P}$ and is therefore **excluded**.

Note that here, unlike in Chapter 6, the homogeneizing coordinate y_0 is put first according to its index since no coordinate of a projective point is distinguished among the others. For example, each of the $n + 1$ equations $y_0 = 0, \ldots, y_n = 0$ represents a projective hyperplane in $\mathcal{P}$. The plane $y_0 = 0$ is singled out only by its affine interpretation above as the **ideal plane** $\mathcal{A}_\infty$ of $\mathcal{A}$, while the other planes correspond to the coordinate planes $x_1 = 0, \ldots, x_n = 0$ in $\mathcal{A}$. The points of $\mathcal{A}_\infty$ are the **ideal points**. They correspond to the directions of $\mathcal{A}$. The union of $\mathcal{A}$ and $\mathcal{A}_\infty$ is called the **projective extension** of $\mathcal{A}$, and it forms the n-dimensional projective space $\mathcal{P}$. Since the coordinate y_0 is superfluous within the ideal plane, $\mathcal{A}_\infty$ itself forms a projective space of dimension $n - 1$.

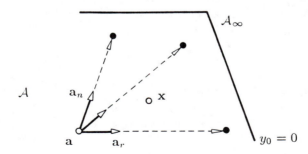

Figure 20.1: The projective extension of an affine space.

Notation: Throughout this part of the book small hollow letters denote homogeneous coordinate columns of a point and also the point itself. Thus, for example, the coordinate columns x and $y = x\varrho$ represent the same point denoted also by x or y. Similar to linear and affine spaces, $\mathcal{P}^n$ denotes the n-dimensional projective space obtained from $\mathcal{A}^n$.

Remark 1: Given an n-dimensional projective space $\mathcal{P}$ one can view any hyperplane $\mathcal{H}$ as its ideal plane and $\mathcal{P} \setminus \mathcal{H}$ as the corresponding affine space. Then $\mathcal{P}$ is the projective extension of $\mathcal{P} \setminus \mathcal{H}$. In particular, if $\mathcal{H}$ is given by $y_0 = 0$, the points of $\mathcal{P} \setminus \mathcal{H}$ have the affine coordinates $y_i/y_0, i = 1, \ldots, n$.

20.2 Projective Coordinates

Any $r + 1$ points $\mathbb{p}_0, \ldots, \mathbb{p}_r$ of a projective space $\mathcal{P}$ are said to be projectively independent if their $r + 1$ coordinate columns $\mathbb{p}_i$ are linearly independent. The **span** of $r + 1$ projectively independent points $\mathbb{p}_0, \ldots, \mathbb{p}_r$ in $\mathcal{P}$ consists of all points represented by a linear combination of the columns $\mathbb{p}_i$,

$$\mathbb{p}\varrho = \mathbb{p}_0 x_0 + \mathbb{p}_1 x_1 + \cdots + \mathbb{p}_r x_r .$$

The coefficients x_i are not determined by the points $\mathbb{p}_0, \ldots, \mathbb{p}_r$ and $\mathbb{p}$ since the coordinates of the points $\mathbb{p}_i$ are only defined up to some individual factor. In order to get unique coefficients in the representation above one distinguishes a further reference point $\mathbb{p}_u$ not in the span of any proper subset of $\{\mathbb{p}_0, \ldots, \mathbb{p}_r\}$. Then one chooses the coordinates $\mathbb{p}_i$ such that

$$\mathbb{p}_u \varrho = \mathbb{p}_0 + \cdots + \mathbb{p}_r .$$

Picking a unit point fixes the coordinates of $\mathbb{p}_0, \ldots, \mathbb{p}_r$ up to the common factor ϱ and consequently fixes the coefficients x_i or, more exactly, the ratios of the x_i of an arbitrary point $\mathbb{p}$. Then the x_i are called the **projective coordinates** of $\mathbb{p}$ with respect to the **frame** $\mathbb{p}_0, \ldots, \mathbb{p}_r; \mathbb{p}_u$, while $\mathbb{p}_u$ is referred to as the **unit point**.

Since the span of $\mathbb{p}_0, \ldots, \mathbb{p}_r$ can be described by projective, i.e., homogeneous coordinates, the span is an r-dimensional projective subspace of $\mathcal{P}$. Note that $\mathbb{p}_0, \ldots, \mathbb{p}_r, \mathbb{p}_u$ are projectively dependent, while any $r + 1$ of these points are independent and span the same subspace.

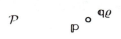

Example 1: Any two projectively dependent points in $\mathcal{P}^n$ satisfy $\mathbb{p} = \mathbb{q}\varrho$. This implies that these points are identical.

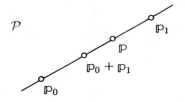

Example 2: A projective line in $\mathcal{P}^n$ is spanned by two independent points $\mathbb{p}_0$ and $\mathbb{p}_1$. On choosing a unit point $\mathbb{p}_u = \mathbb{p}_0 + \mathbb{p}_1$ each point on the line can be written as

$$\mathbb{p}\varrho = \mathbb{p}_0 x_0 + \mathbb{p}_1 x_1 \ ,$$

where x_0, x_1 are the (homogeneous) projective coordinates of $\mathbb{p}$ with respect to $\mathbb{p}_0$, $\mathbb{p}_1$ and $\mathbb{p}_u$. The inhomogeneous coordinates $x = x_1/x_0$ define a **projective scale** on the line.

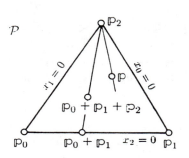

Example 3: A projective plane in $\mathcal{P}^n$ is spanned by three independent points $\mathbb{p}_0$, $\mathbb{p}_1$ and $\mathbb{p}_2$. On choosing a unit point $\mathbb{p}_u = \mathbb{p}_0 + \mathbb{p}_1 + \mathbb{p}_2$, each point of the plane can be written as

$$\mathbb{p}\varrho = \mathbb{p}_0 x_0 + \mathbb{p}_1 x_1 + \mathbb{p}_2 x_2 \ ,$$

where x_0, x_1, x_2 are the (homogeneous) projective coordinates of $\mathbb{p}$ with respect to $\mathbb{p}_0$, $\mathbb{p}_1$, $\mathbb{p}_2$ and $\mathbb{p}_u$.

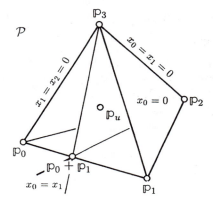

Example 4: A projective 3-space in $\mathcal{P}^n$ is spanned by four independent points $\mathbb{p}_0, \ldots, \mathbb{p}_3$. On choosing a unit point $\mathbb{p}_u = \mathbb{p}_0 + \cdots + \mathbb{p}_4$, each point of the 3-space can be written as

$$\mathbb{p}\varrho = \mathbb{p}_0 x_0 + \cdots + \mathbb{p}_3 x_3 \ ,$$

where $x_0, \ldots, x_3$ are (homogeneous) projective coordinates of $\mathbb{p}$ with respect to $\mathbb{p}_0, \ldots, \mathbb{p}_3$ and $\mathbb{p}_u$.

In particular, if $r = n$, the representation

$$\mathbb{p}\varrho = \mathbb{p}_0 x_0 + \cdots + \mathbb{p}_n x_n$$

together with the unit point $\mathbb{p}_u = \mathbb{p}_0 + \cdots + \mathbb{p}_n$ defines the **new projective coordinates** in $\mathcal{P}$ with respect to the new frame $\mathbb{p}_0, \ldots, \mathbb{p}_n; \mathbb{p}_u$.

20.3   The Equations of Planes and Subspaces

A **projective hyperplane** of $\mathcal{P}^n$ is an $n-1$-dimensional subspace, i.e., the span of n independent points in $\mathcal{P}^n$. Hence, the solution of some linear equation

$$\mathbb{u}^t \mathbb{x} = u_0 x_0 + u_1 x_1 + \cdots + u_n x_n = 0, \quad \mathbb{u} \neq \mathbb{o},$$

represents a projective hyperplane and, conversely, each projective hyperplane has such an equation. The homogeneous coefficients u_i are called projective **hyperplane coordinates**. An inhomogeneization shows that a projective hyperplane can be viewed as the union of an affine hyperplane in $\mathcal{P}^n$ given by

$$\begin{bmatrix} u_1 & \ldots & u_n \end{bmatrix} \frac{\mathbf{x}}{x_0} + u_0 = 0$$

with its ideal points $\mathbf{v}$ represented by the solution of $\begin{bmatrix} u_1 & \ldots & u_n \end{bmatrix} \mathbf{v} = 0$.

Let $\mathcal{S}$ be the intersection of $s + 1$ hyperplanes given by $\mathbb{u}_0, \ldots, \mathbb{u}_s$. Then $\mathcal{S}$ forms the solution of

$$\begin{bmatrix} \mathbb{u}_0 & \ldots & \mathbb{u}_s \end{bmatrix}^t \mathbb{p} = \mathbb{o}$$

which can be computed by the Gauss–Jordan algorithm. If the planes are independent, i.e., if $u_0, \ldots, u_s$ are linearly independent, the solution S is spanned by $n - s$ independent points $p_{s+1}, \ldots, p_n$. One has

$$[u_0 \ \ldots \ u_s]^t \left[p_{s+1} \ \cdots \ p_n \right] = 0 \ ,$$

or in blocks

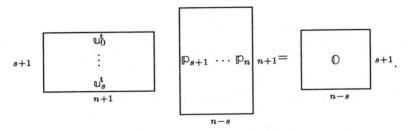

Conversely, one can compute the u's from any given p's. This means that any projective subspace of dimension r can be represented by the intersection of $n - r$ hyperplanes. The affine counterpart to this statement is discussed in Section 9.3.

20.4 The Equation of a Point

The linear expression

$$u^t x = u_0 x_0 + u_1 x_1 + \cdots + u_n x_n = 0$$

is symmetric in u and x. If $u \neq o$ is given, this expression represents the **equation of a hyperplane**: All points x such that $u^t x = 0$ form a hyperplane which has the homogeneous coordinates u. On the other hand, if $x \neq o$ is given, $u^t x = 0$ represents the **equation of a point**: All planes with coordinates u such that $u^t x = 0$ contains the point x. In correspondence with Example 3, Figure 20.2 shows the equations of a few special points in the plane.

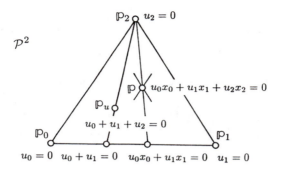

Figure 20.2: The equations of some points.

The intersection of n independent hyperplanes u_i defines a point x. This point x can be computed as the solution of the linear system $u_i^t x = 0$, $i = 1, \ldots, n$. On the other hand, n projectively independent points x_i span a hyperplane u which can be computed as the solution of the linear system $x_i^t u = 0, i = 1, \ldots, n$. Both systems differ only in the notation, i.e., there is no difference computationally between the intersection of n hyperplanes and the span of n points.

The equation of the common point of the hyperplanes $u_1, \ldots, u_n$ can be written as

$$\det [u \; u_1 \; \ldots \; u_n] = 0 \ .$$

Expanding the determinant along the first column one gets $u^t x = 0$, where the coordinates of x are the cofactors. Analogously, the equation of the hyperplane spanned by $x_1, \ldots, x_n$ can be written as

$$\det [x \; x_1 \; \ldots \; x_n] = 0 \ .$$

Expanding the determinant along the first column one gets $u^t x = 0$, where the coordinates of u are now the cofactors.

Example 5: Two points a and b in the plane define a line. One gets

$$u_i = \varrho \det \begin{bmatrix} a_j & b_j \\ a_k & b_k \end{bmatrix} \ , \qquad i,j,k = 0,1,2 \text{ past round.}$$

Example 6: Two lines u and v in the plane define a point p. One gets

$$p_i = \varrho \det \begin{bmatrix} u_j & v_j \\ u_k & v_k \end{bmatrix} \ , \qquad i,j,k = 0,1,2 \text{ past round.}$$

20.5 Pencils and Bundles

Let u_0 and u_1 be the homogeneous coordinates of two independent hyperplanes. Then the planes given by

$$u = u_0 y_0 + u_1 y_1$$

form a **pencil** of hyperplanes, all of which contain the intersection $\mathcal{B}$ of the planes u_0 and u_1, as illustrated in Figure 20.3. $\mathcal{B}$ is called the **base** or **center** of the pencil.

A comparison with Example 2 reveals that the planes of the pencil form a one-dimensional projective space. Moreover, y_0 and y_1 are the projective coordinates of the plane u with respect to the frame u_0, u_1 and $u_0 + u_1$, and $y = y_1/y_0$ defines a projective scale.

This projective scale can be carried over to any line disjoint from $\mathcal{B}$ by the incident relation $u^t x = 0$, where u is a plane of the pencil and x is a point on the line. Namely, let u_i and x_i be chosen such that $u_0^t x_0 = 0$, $u_1^t x_1 = 0$, and $[u_0 + u_1]^t [x_0 + x_1] = u_0^t x_1 + u_1^t x_0 = 0$; then $x = x_0 y_0 + x_1 y_1$ lies on the plane $u_0 y_0 + u_1 y_1$ which has the same projective coordinates y_i.

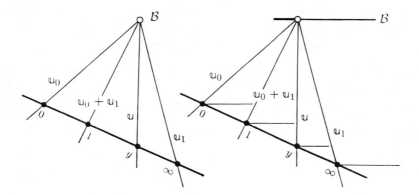

Figure 20.3: A pencil of lines in the plane and a pencil of planes in the space.

Let u_0, u_1, u_2 be the homogeneous coordinates of three independent hyper-planes. Then the planes given by

$$u = u_0 y_0 + u_1 y_1 + u_2 y_2$$

form a **bundle** of planes, where all planes contain the intersection $\mathcal{B}$ of the planes u_0, u_1 and u_2, as illustrated in the left part of Figure 20.4. $\mathcal{B}$ is called the **base** or **center** of the bundle. Note that $dim\,\mathcal{B} = n - 3$.

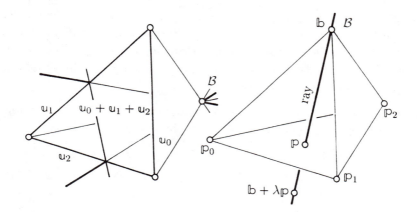

Figure 20.4: A bundle of planes and a bundle of rays in 3-space.

A comparison with Example 3 reveals that the planes of the bundle form a two-dimensional projective space. Moreover, y_0, y_1, and y_2 are the projective coordinates of u with respect to the frame $u_0, u_1, u_2; u_0 + u_1 + u_2$.

Any two independent hyperplanes of the bundle define a pencil. The bases of all these pencils form a **bundle of "rays,"** where each ray is incident to exactly one point of any fixed two-dimensional plane which does not intersect B. This is also illustrated in Figure 20.4.

20.6 Duality

The hyperplanes of an n-dimensional projective space $\mathcal{P}$ can be described by homogeneous coordinates. Therefore the hyperplanes form an n-dimensional projective space, called the **dual space** of $\mathcal{P}$, which is denoted by $\mathcal{P}^*$. The symmetry of $u^t x = 0$ in u and x implies that one can view $\mathcal{P}$ as the dual space of $\mathcal{P}^*$.

In terms of coordinates, there is no difference between $\mathcal{P}$ and $\mathcal{P}^*$. In particular, the join B of r independent points corresponds to the join B^* of r hyperplanes. However, one usually interprets B^* as a point set S in $\mathcal{P}$, namely as the intersection of the r hyperplanes. Note that $dim\, B = dim\, B^* = r - 1$ but $dim\, S = n - r$.

Two configurations which only differ in the interpretation of the u's and x's are called **dual**. Note that intersections and joins are dual. An important consequence of this concept is that one can translate each configuration of points and planes in $\mathcal{P}$, plus pertaining statements about joins and intersections, into the dual configuration of planes and points and the pertaining dual statements about intersections and joins.

Examples of simple dual configurations in the plane are shown in Figure 20.5. Simple examples in 3-space are shown in Figure 20.6.

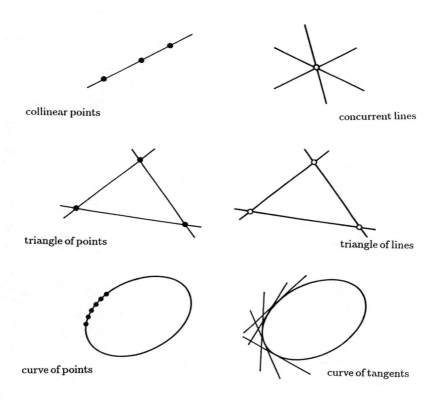

collinear points

concurrent lines

triangle of points

triangle of lines

curve of points

curve of tangents

Figure 20.5: Some dual figures in the plane.

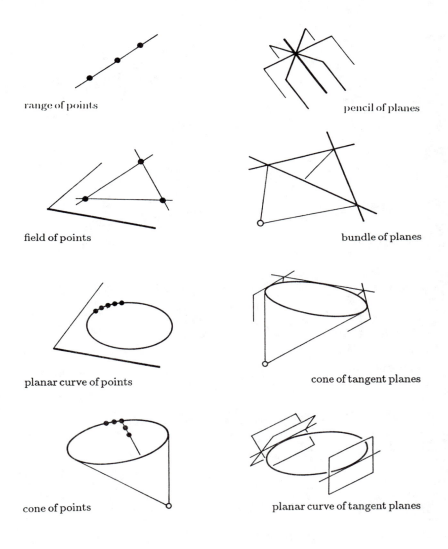

range of points

pencil of planes

field of points

bundle of planes

planar curve of points

cone of tangent planes

cone of points

planar curve of tangent planes

Figure 20.6: Some dual figures in space.

20.7 Notes and Problems

1 The intersection $\mathcal{U}_1 \sqcap \mathcal{U}_2$ and the join $\mathcal{U}_1 \sqcup \mathcal{U}_2$ of two projective sub-spaces satisfy

$$dim\,\mathcal{U}_1 \sqcap \mathcal{U}_2 + dim\,\mathcal{U}_1 \sqcup \mathcal{U}_2 = dim\,\mathcal{U}_1 + dim\,\mathcal{U}_2 \ .$$

2 The planes $\mathsf{u} = \mathsf{u}_0 + \mathsf{u}_1 y$ form a parallel pencil in $\mathcal{A}^n$ if u_1 represents the ideal hyperplane of $\mathcal{A}^n$, i.e., if $\mathsf{u}_1 = [1 \quad 0 \ldots 0]^t$.

3 An inhomogeneization leads to barycentric coordinates. Namely, let a subspace in $\mathcal{P}^n$ be given by

$$\mathsf{x} = \mathsf{a}\alpha + \mathsf{b}\beta + \cdots + \mathsf{z}\zeta \ ,$$

where $\mathsf{x} = \begin{bmatrix} x_0 \\ \mathbf{x} \end{bmatrix}$, $\mathsf{a} = \begin{bmatrix} a_0 \\ \mathbf{a} \end{bmatrix}, \ldots, \mathsf{z} = \begin{bmatrix} z_0 \\ \mathbf{z} \end{bmatrix}$. Then one has

$$\frac{\mathbf{x}}{x_0} = \frac{(\mathbf{a}/a_0)a_0\alpha + \cdots + (\mathbf{z}/z_0)z_0\zeta}{a_0\alpha + \cdots + z_0\zeta}$$

which is the representation of $\mathbf{x}/x_0$ by barycentric coordinates with respect to $\mathbf{a}/a_0, \ldots, \mathbf{z}/z_0$. An example is given in Section 7.6(3).

4 Conversely, consider a barycentric combination

$$\mathbf{x} = \mathbf{a}\alpha + \cdots + \mathbf{z}\zeta \ .$$

Dispensing with the condition $\alpha + \cdots + \zeta = 1$ one gets

$$\mathbf{x}\varrho = \mathbf{a}\alpha + \cdots + \mathbf{z}\zeta \ ,$$

where $\varrho = \alpha + \cdots + \zeta$. Furthermore, one can introduce the **"weights"** $a, \ldots, z$ in order to get

$$\mathbf{x}\varrho = \mathbf{a}a \cdot \frac{\alpha}{a} + \cdots + \mathbf{z}z \cdot \frac{\zeta}{z} \ .$$

Then $\alpha/a, \ldots, \zeta/z$ are the projective coordinates of $\mathbf{x}$ with respect to the basis points $\mathbf{a}, \ldots, \mathbf{z}$, weighted by $a, \ldots, z$, and the unit point

$(\mathbf{a}a + \cdots + \mathbf{z}z)/(a + \cdots + z)$. Recall that $\mathbf{x}, \mathbf{a}, \ldots, \mathbf{z}$ can be represented by affine, extended, or barycentric coordinates.

5 In particular, one can consider new projective coordinates on a projective line

$$\begin{bmatrix} x_0 \\ x_1 \end{bmatrix} = \begin{bmatrix} a_0 \\ a_1 \end{bmatrix} y_0 + \begin{bmatrix} b_0 \\ b_1 \end{bmatrix} y_1 \ .$$

Inhomogeneizing gives

$$x = \frac{x_1}{x_0} = \frac{a_1 + b_1 y}{a_0 + b_0 y}, \quad y = \frac{y_1}{y_0} \ ,$$

which corresponds to a **linear rational transformation** of the associated projective scales.

6 Dual configurations are sometimes called **reciprocal** configurations, and dual theorems are sometimes called **reciprocal** theorems.

21 Projective Maps

A projective map maps a projective space into another projective space while preserving the projective structure. Consequently, a projective mapping maps any configuration defined by projective coordinates onto a configuration which is defined by the same coordinates with respect to the image of the coordinate frame. Special considerations are necessary if the image degenerates.

Literature: Samuel, Stolfi, Wylie

21.1 Matrix Notation

Let $\Phi : \mathcal{P}^m \to \mathcal{P}^n$ be a map, let $\mathfrak{q}_0, \ldots, \mathfrak{q}_m, \mathfrak{q}_u$ be the image of the coordinate frame
$\mathbb{p}_0, \ldots, \mathbb{p}_m; \mathbb{p}_u$ of $\mathcal{P}^m$, and let

$$\mathfrak{q}_u = \mathfrak{q}_o \sigma_0 + \cdots + \mathfrak{q}_m \sigma_m .$$

Then, Φ is a **projective map** if $\varkappa \varrho = \mathbb{p}_0 x_0 + \cdots + \mathbb{p}_m x_m$ is mapped onto $\mathsf{y}\sigma = \mathfrak{q}_0 \sigma_0 x_0 + \cdots + \mathfrak{q}_m \sigma_m x_m$. In matrix notation this projective map $\Phi : \varkappa \mapsto \mathsf{y}$ is written as

$$\mathsf{y}\sigma = \mathbb{P}\varkappa = [\mathfrak{q}_0 \sigma_0 \ \ldots \ \mathfrak{q}_m \sigma_m]\varkappa .$$

Note that some of the $\mathfrak{q}$'s or y can be allowed to be $\mathfrak{o}$, which means that Φ is allowed to map into $\mathcal{P}^n \cup \{\mathfrak{o}\}$. However, if $\mathfrak{q}_i \neq \mathfrak{o}$ one has to require that $\sigma_i \neq 0$ since the i-th column of $\mathbb{P}$ must represent the point $\mathfrak{q}_i$.

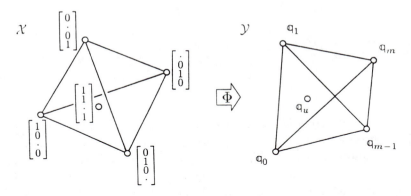

Figure 21.1: Fundamental points and projective mapping.

One should convince oneself that a projective map maps projective sub-spaces onto projective subspaces, possibly including $\odot$, and that its restriction onto a subspace is a projective map again.

Example 1: Consider the projective map illustrated in Figure 21.2. By inspection one gets

$$\mathbb{P} = \begin{bmatrix} \alpha & \beta & 0 \\ 0 & 0 & 1 \end{bmatrix}, \qquad \text{where} \quad \alpha + \beta = 1 \quad \text{and} \quad \alpha, \beta \neq 0 .$$

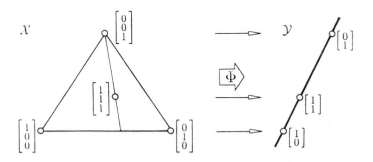

Figure 21.2: Example of a projective map.

Remark 1: An affine map $\mathbf{y} = \mathbf{a} + A\mathbf{x}$ can be represented by homogeneous coordinates,

$$\mathbf{y} = \begin{bmatrix} 1 & \mathbf{o}^t \\ \mathbf{a} & A \end{bmatrix} \mathbf{x} \; .$$

Since an ideal point $\begin{bmatrix} 0 \\ \mathbf{x} \end{bmatrix}$ is mapped onto an ideal point $\begin{bmatrix} 0 \\ \mathbf{y} \end{bmatrix}$, the ideal plane is fixed under an affine map.

Remark 2: Regarding a projective space as the set of all one-dimensional subspaces of a linear space means that a projective map induces a linear map and vice versa.

21.2 Exceptional Spaces

If the projective map Φ is not a one-to-one mapping, it maps a special subspace $\mathcal{Z}$ of $\mathcal{P}^m$ onto the excluded "point" $\mathbf{o}$. In this case $\mathbf{o}$ has to be added to the image space. Then one has for all $\mathbf{z} \in \mathcal{Z}$

$$\mathbb{P}\mathbf{z} = \mathbf{o} \; .$$

The subspace $\mathcal{Z}$ is called the **exceptional space** or **null-space** of Φ. It has the property that for any point $\mathbf{x} \in \mathcal{P}^m$ the span of $\mathbf{x}$ and $\mathcal{Z}$ is mapped onto the image of $\mathbf{x}$,

$$\mathbb{P}\left[\mathbf{x} + \lambda \mathbf{z}\right] = \mathbb{P}\mathbf{x} + \mathbf{o} = \mathbb{P}\mathbf{x} \; .$$

The space spanned by $\mathbf{x}$ and $\mathcal{Z}$ is called the **fiber** of $\mathbf{x}$, and one has $rank\,\mathbb{P} + dim\,\mathcal{Z} = m$.

An exceptional space exists if the linear system

$$\mathbb{Q}\mathbf{s} = \mathbf{q}_0 \sigma_0 + \cdots \mathbf{q}_m \sigma_m = \mathbf{q}_u$$

has more than one solution, i.e., if $rank\,\mathbb{Q} \leq m$. Then the map Φ is in general not uniquely defined by $\mathbf{q}_0, \ldots, \mathbf{q}_m, \mathbf{q}_u$. However, if the base points $\mathbb{P}_0, \ldots, \mathbb{P}_r$ span $\mathcal{Z}$, i.e., if

$$\mathbb{Q} = \begin{bmatrix} \mathbf{o} & \cdots & \mathbf{o} & \mathbf{q}_{r+1} & \cdots & \mathbf{q}_m \end{bmatrix}$$

with $q_{r+1}, \ldots, q_m$ being independent, then the factors σ_i in

$$q_{r+1}\sigma_{r+1} + \cdots + q_m\sigma_m = q_u$$

are unique and so is Φ. See Figure 21.3.

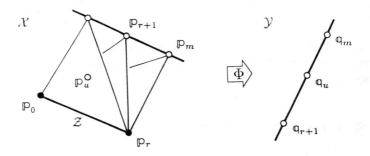

Figure 21.3: The exceptional space $\mathcal{Z}$ of Φ.

Example 2: In Section 7.2 the exceptional space consists of the eye while the fibers are the projection rays.

Example 3: The map constructed in Example 1 is not uniquely specified. Different exceptional are spaces possible. They are given by the points $\mathbf{z} = [-\beta \ \alpha \ 0]^t$.

Example 4: Particular projective maps are considered in Chapters 6, 7, and 8. These maps are **idempotent**, i.e., one has $\Phi \circ \Phi = \Phi$.

Remark 3: In case Φ has a non-empty exceptional space, all subspaces of $\mathcal{Y}$ contain $\mathbf{o}$.

21.3 The Dual Map

Let $\mathcal{X}$ and $\mathcal{Y}$ be projective spaces, and let $\Phi : \mathcal{X} \to \mathcal{Y}$ be a projective map given by

$$\mathbf{y}\sigma = \mathbb{P}\mathbf{x} .$$

Furthermore, let the image y of a point x lie in a hyperplane v; then x lies in a hyperplane u since $v^t y \sigma = v^t \mathbb{P} x = \varrho u^t x$, where

$$u\varrho = \mathbb{P}^t v \ .$$

This relation between v and u establishes a projective map $\Phi^* : \mathcal{Y}^* \to \mathcal{X}^*$. Φ^* is called the **dual map** of Φ. If, however, v contains the image of $\mathcal{X}$, one has $u = o$, i.e., v belongs to the exceptional space of Φ^*, and o must be added to $\mathcal{X}^*$. Note that the image u of any hyperplane v contains the exceptional space $\mathcal{Z}$ since for any point $z \in \mathcal{Z}$

$$\varrho u^t z = v^t \mathbb{P} z = v^t o = 0 \ .$$

Note that the dual map Φ^{**} of Φ^* coincides with Φ.

Example 5: A useful application is described in Section 8.2.

21.4 Collineations and Correlations

An invertible projective map $\Phi : \mathcal{X} \to \mathcal{X}$ is termed a **collineation** because Φ maps any three points of each line onto three points of a line.

However, if Φ is invertible and maps $\mathcal{X}$ onto the dual space $\mathcal{X}^*$ it is called a **correlation**. Any three points of a line are mapped onto three hyperplanes which intersect along a line.

There is no difference between a projective space $\mathcal{X}$ and its dual space $\mathcal{X}^*$ besides the terminology. Hence, the dual map of a collineation is a collineation since it maps any three "collinear" hyperplanes into three "collinear" hyperplanes. The dual map of a correlation $v\sigma = \mathbb{P} x$ is a correlation $u\varrho = \mathbb{P}^t y$ since it maps collinear points into collinear hyperplanes.

Remark 4: More generally, one can define a collineation as an invertible map of a projective space onto itself which maps lines onto lines. See Problem 5 in Section 21.7.

Remark 5: Any collineation of $\mathcal{P}^n$ onto itself can be realized by an embedding of $\mathcal{P}^n$ into $\mathcal{P}^{n+1}$ and a sequence of projections. Recall that Φ is defined by $n + 2$ points and their images. Figure 21.4 illustrates this principle for two lines in the plane.

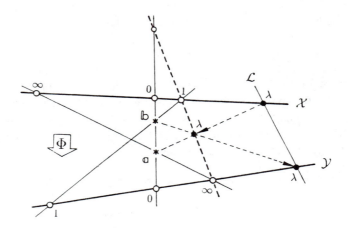

Figure 21.4: Composing a projectivity from two perspectivities.

Example 6: A nice example of a correlation used in nomography was given by Adam. His mechanical scanner consists of a circular disk pivoting on a rectangular plate which glides along a fixed ruler. Disk and plate have several lines engraved as indicated in Figure 21.5. The scanner realizes the correlation $\Phi : \varkappa \mapsto \mathsf{v}$ given by

$$\mathsf{v}\sigma = \begin{bmatrix} b & 1 & \\ -1 & & \\ & & 1 \end{bmatrix} \begin{bmatrix} 1 \\ \mathbf{x} \end{bmatrix},$$

and the inverse of the correspondingdual map Φ^*

$$\begin{bmatrix} 1 \\ \mathbf{y} \end{bmatrix} \varrho = \begin{bmatrix} & 1 & \\ -1 & b & \\ & & 1 \end{bmatrix} \mathsf{u}$$

by turning the scanner upside down.

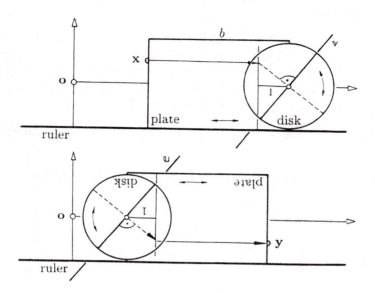

Figure 21.5: Adam's scanner.

Example 7: If a correlation coincides with its dual map, one has $\mathbb{P} = \mathbb{P}^t$. Then the correlation is a polarity with respect to a quadric, cf. Chapter 23.

21.5 The Crossratio

Consider four points x, y, a, and b of a line corresponding to the values ξ, η, α, and β, respectively, of a projective scale. Then the value

$$\delta = \frac{\xi - \alpha}{\xi - \beta} : \frac{\eta - \alpha}{\eta - \beta}$$

is called the **crossratio** of x and y with respect to a and b, denoted more concisely by

$$\delta = cr\,[xy|ab] \ .$$

It has the following important property which was already known by Pappus(< 300 A. D.):

The crossratio of four points does not depend on the projective
scale used.

In other words, δ is invariant under projective transformations. For a proof
let α, β, η be fixed. Then δ is a rational linear polynomial in ξ, and it
forms a projective scale, see Section 20.7(5). Since $\xi = \alpha, \eta, \beta$ corresponds
to $\delta = 0, 1, \infty$, respectively, the crossratio δ is the unique projective scale
defined by the frame ɑ, y, ʙ. This proves the theorem above.

Note that the fourth point ϰ is uniquely defined by the other three points
and δ.

Remark 6: By homogeneizing the projective scale one obtains

$$\delta = \frac{\det [\boldsymbol{\xi} \quad \boldsymbol{\alpha}]}{\det [\boldsymbol{\xi} \quad \boldsymbol{\beta}]} : \frac{\det [\boldsymbol{\eta} \quad \boldsymbol{\alpha}]}{\det [\boldsymbol{\eta} \quad \boldsymbol{\beta}]} \, ,$$

where $\boldsymbol{\xi} = [\xi_0 \quad \xi_1]^t$, $\xi = \xi_1/\xi_0$, etc.

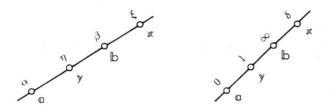

Figure 21.6: Crossratios of four points.

Remark 7: From Section 20.5 it follows that any four members of a pencil
of hyperplanes and their intersections with a straight line define the same
crossratio. As an application one can compute a crossratio in a pencil by
means of an affine scale. This is illustrated in Figure 21.7, where α and β
are chosen to be 0 and ∞, respectively. It follows that $cr\,[\varkappa y|\text{ɑʙ}] = \xi/\eta$.

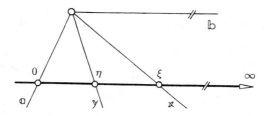

Figure 21.7: Crossratio and affine scale.

21.6 Harmonic Position

Of particular interest are points corresponding to the special crossratio $\delta = -1$, i.e., where

$$2(\xi\eta + \alpha\beta) = (\xi + \eta)(\alpha + \beta) \ .$$

Obviously, interchanging ξ and η, or α and β, or the pairs α, β and ξ, η does not change the crossratio $\delta = -1$. This is why x, y are said to be in **harmonic position** with respect to a, b .

Assuming an affine scale as in Remark 7, two points x and y are in harmonic position with respect to a and b, where $\alpha = 0$ and $\beta = \infty$, if and only if $\xi = -\eta$, i.e., if a is the midpoint of x and y, as illustrated in Figure 21.8. This fact can easily be used to construct the "fourth harmonic point."

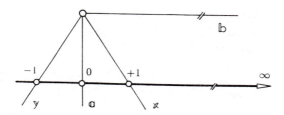

Figure 21.8: Harmonic position.

21.7 Notes and Problems

1 A collineation $\mathbb{P}$ which equals its inverse is called an **involution**, i.e., one has $\mathbb{P}\mathbb{P} = \varrho\mathbb{I}$, where $\mathbb{I}$ represents the identity.

2 A projective map $\mathbb{P}$ with the property $\mathbb{P}\mathbb{P} = \varrho\mathbb{P}$ is called a **projection**.

3 In the field of computer graphics, it is common practice to decompose a perspective mapping $\mathcal{P}^3 \to \mathcal{P}^2$ into a projective map $\mathcal{P}^3 \to \mathcal{P}^3$ and a subsequent parallel projection $\mathcal{P}^3 \to \mathcal{P}^2$ which is an affine map. This simplifies visibility computations since the depth of geometric objects like planes can be linearly interpolated in the intermediate image, the spatial perspective of Section 6.7.

4 Two transformations such as $y\sigma = \mathbb{P}x$ and $u\varrho = \mathbb{P}^t v$ are called **contragredient**.

5 Let $\bar{\mathbb{p}}$ denote the coordinate column obtained from $\mathbb{p}$ by taking the complex conjugate numbers. Show that $\Phi : \mathcal{P}^n \to \mathcal{P}^n$, $\Phi(\mathbb{p}) = \bar{\mathbb{p}}$, is a collineation of a complex projective space but not a projective map.

6 The map defined in Note 5 leaves crossratios unchanged. It is the only non-projective map with this property (Staudt 1847).

7 The crossratio of four imaginary points of a line is real if the four corresponding imaginary values of some projective scale lie on a circle of the complex plane and vice versa.

8 If $cr[xy|ab] = \delta$ then

$$cr[xy|ba] = cr[yx|ab] = 1/\delta,$$
$$cr[xa|yb] = cr[by|ax] = 1 - \delta,$$
$$cr[xb|ay] = cr[ay|xb] = \delta/(\delta - 1).$$

9 The 24 permutations of x, y, a, b in $cr[xy|ab] = \delta$ lead to only 6 different values of the crossratio, namely

$$\delta, \quad \frac{1}{\delta}, \quad 1 - \delta, \quad \frac{1}{1-\delta}, \quad \frac{\delta}{\delta-1}, \quad \frac{\delta-1}{\delta}.$$

10 The points ξ and η are in harmonic position with respect to α and $\beta = -\alpha$ if $\xi \cdot \eta = \alpha^2$.

22 Some Projective Figures

Projective mappings preserve the structure of projective spaces. Moreover, there are properties of figures which are invariant under projective mappings. These properties depend on some construction rules, e.g., the rules for drawing a fourth harmonic point, but they do not depend on the specific shape and position of the figure in a projective space. Often the use of this invariance this can simplify proofs.

Literature: Berger, Blaschke, Samuel, Wylie

22.1 Complete Quadruples in the Plane

Two of the simplest but nevertheless interesting figures in the plane are the complete quadrangle and its dual configuration, the complete quadrilateral.

The **complete quadrangle** is defined by four points which are marked by solid dots, •, in Figure 22.1. All possible pairs of points are connected by three pairs of opposite sides which meet in three diagonal points marked by ∘. A diagonal meets the complete quadrangle in two pairs of points, namely two diagonal points and the intersection with two opposite sides. One has:

Both pairs are in harmonic position.

This can be observed from the fact that the quadrangle can be seen as the projective image of the unit square in the (projectively extended) affine plane.

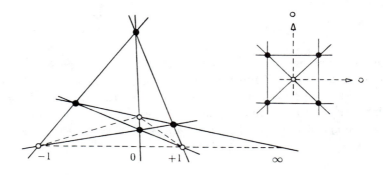

Figure 22.1: The complete quadrangle.

The **complete quadrilateral** is defined by four lines in the plane, as illustrated in Figure 22.2. There are three pairs of opposite vertices which define the three (dashed) diagonals. Connecting a diagonal point of the complete quadrilateral with the vertices gives two pairs of lines, namely two diagonals and the lines through the two remaining opposite vertices. One has

Both pairs are in harmonic position.

The quadrilateral is dual to the complete quadrangle and can be viewed as a projective image of the four sides of the unit square in the (projectively extended) affine plane.

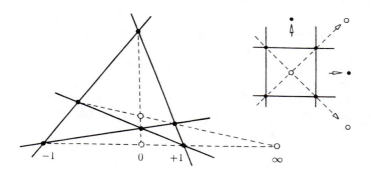

Figure 22.2: The complete quadrilateral.

The projective image of a regular quadrangular lattice, as depicted in Figure 22.3, is called a **Möbius net**. It is determined by the image of only one quadrangle from which it can be completed with the aid of diagonals, just like a regular lattice in the affine plane. By adding diagonals, one can obtain the image of a **regular triangular net** and also that of a **honeycomb net**.

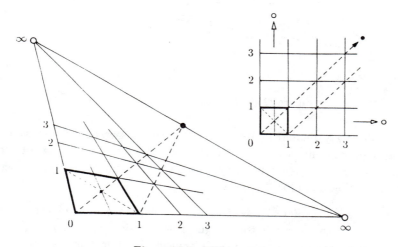

Figure 22.3: Möbius net.

22.2 Desargues' Configuration

Consider two triangles formed by $\mathfrak{a}$, $\mathfrak{b}$, $\mathfrak{c}$ and by $z + \mathfrak{a}$, $z + \mathfrak{b}$, $z + \mathfrak{c}$, as illustrated in Figure 22.4. The corresponding sides of the two triangles meet in points of a line $\mathcal{W}$. This can be seen as follows: The lines through $\mathfrak{a}$, $\mathfrak{b}$ and $z + \mathfrak{a}$, $z + \mathfrak{b}$ intersect in $\mathfrak{a} - \mathfrak{b}$, etc. The three intersection points lie on a line since $(\mathfrak{a} - \mathfrak{b}) + (\mathfrak{c} - \mathfrak{a}) + (\mathfrak{b} - \mathfrak{c}) = \mathfrak{o}$. Note that the point $z + \mathfrak{a}$ can lie anywhere on the line through $z, \mathfrak{a}$ depending on the multiple used to represent the point $\mathfrak{a}$.

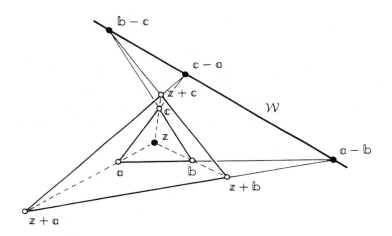

Figure 22.4: Desargues' configuration.

Two projectively related figures are said to be in **perspective position** if corresponding points lie on lines through some fixed point or if corresponding lines intersect in some fixed hyperplane. Employing this terminology, **Desargues' theorem** (1639) takes on the form:

> If the vertices of two triangles in the plane are in perspective position with respect to some point, then the sides of the triangles are in perspective position with respect to some line and vice versa.

Remark 1: Evidently, the dual of Desargues' configuration happens to exhibit the inverse theorem.

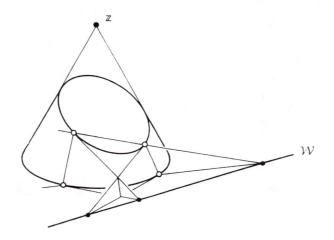

Figure 22.5: Perspective ellipses.

Remark 2: As a consequence of Desargues' theorem, every collineation in $\mathcal{P}^n$ which maps each line through some fixed point onto itself also leaves some hyperplane unchanged point by point. Such a collineation is called a **perspectivity**. Figure 22.5 shows an example.

22.3 Pappus' Configuration

Consider a hexagon in the plane whose vertices $1, 2, \ldots, 6$ alternately lie on two lines. Then one has:

The pairs of opposite sides intersect in three collinear points.

This is illustrated in Figure 22.6. The configuration can be seen as the projective image of a special affine configuration where two intersections lie on the ideal line. Then one only needs to show that the third pair of opposite sides is parallel.

The dual configuration consists of a hexagon whose sides pass through two points alternately. The dual form of Pappus' theorem implies that the three diagonals meet in a common point, as illustrated in Figure 22.7. Note the similarity with Figure 22.6; only the notation has changed.

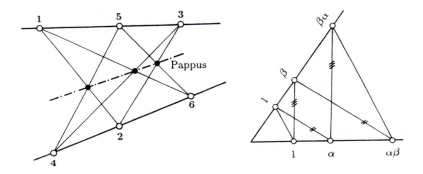

Figure 22.6: Pappus' configuration.

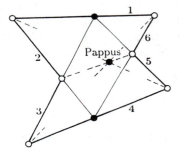

Figure 22.7: The dual configuration.

22.4 Conic Sections

A conic in the projective plane can be viewed as the projective extension of an affine conic and can be simply obtained by the introduction of
homogeneous coordinates. Moreover, the non-degenerate conic sections,
hyperbolas, ellipses, and parabolas, are projectively identical as one can
observe from the planar intersections of a cone.

Consider a point p on a non-degenerate conic section Q and four lines
through p intersecting Q in $x_1, \ldots, x_4$.

> The crossratio δ of the four rays does not depend on the position
> of p on Q for fixed $x_1, \ldots, x_4$.

Moreover, let t represent the intersection of both tangents of Q at x_1 and
x_2.

> The crossratio τ of the four rays from t to $x_1, \ldots, x_4$ equals δ^2.

Both of these facts can be verified as illustrated in Figure 22.8. Let p_1 and
p_2 represent different positions of p, and let b represent the intersection
of both tangents of Q at p_1 and p_2. On choosing a new affine coordinate system where b is the origin, p_1 and p_2 are the ideal points of the
coordinates axes and $[1 \quad 1]^t$ lies on Q, the equation of Q becomes $xy = 1$.

Let $[x_i \quad y_i]^t$ denote the new coordinates of x_i, then the crossratios of the
lines through p_1 and p_2 are

$$cr[x_1 x_2 | x_3 x_4] \quad \text{and} \quad cr[y_1 y_2 | y_3 y_4] ,$$

respectively. Since x and y are related by $y = 1/x$ both crossratios are
equal.

In particular, let $p_1 = x_1$ and $p_2 = x_2$, then $x_1 = 0$, $y_2 = 0$, and the
crossratio is simply $\delta = x_3 y_4$. Moreover, the four rays from t to $x_1, \ldots, x_4$
have the crossratio $\tau = (x_3 y_4)^2 = \delta^2$.

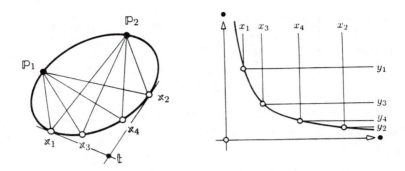

Figure 22.8: Six points on a conic section.

Remark 3: If $\varkappa_1$ varies on C while $\varkappa_2, \varkappa_3, \varkappa_4$ and $\mathbb{p}$ are fixed, δ defines a projective scale on C, i.e., C inherits the projective structure of the pencil with base $\mathbb{p}$. As shown above, this structure does not depend on the position of $\mathbb{p}$ on C.

Remark 4: As a consequence, every non-degenerate conic section Q can be constructed from two projectively related pencils where corresponding lines intersect in points of Q (Steiner 1832).

22.5 Pascal's Theorem

Pappus' configuration can be viewed as a degenerate case of Pascal's theorem (1840) on conic sections which is illustrated in Figure 22.9.

If the vertices $\mathbf{1, 2, \ldots, 6}$ of a hexagon lie on a non-degenerate conic section, then the pairs of opposite sides intersect in points of a straight line.

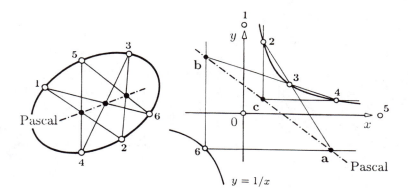

Figure 22.9: Pascal's hexagrammum mysticum.

A simple proof of Pascal's theorem, where Pascal's line does not meet the conic, is given in Section 22.9(9). The general proof makes use of the hyperbola $y = 1/x$ again. Let $[x_i \ \ y_i]^t$ denote the affine coordinates of the vertex $\mathbf{i}$ where $y_i = 1/x_i$. Assuming $x_1 = 0$, $y_5 = 0$ and $x_6 = y_6 = -1$, some elementary calculations show that the points $\mathbf{a}$, $\mathbf{b}$, $\mathbf{c}$ defined by the figure are

$$\mathbf{a} = \begin{bmatrix} x_2 + x_3 + x_2 x_3 \\ -1 \end{bmatrix} \ , \quad \mathbf{b} = \begin{bmatrix} -1 \\ y_3 + y_4 + y_3 y_4 \end{bmatrix} \ , \quad \mathbf{c} = \begin{bmatrix} x_2 \\ y_4 \end{bmatrix} \ .$$

This means that $\mathbf{a}$, $\mathbf{b}$, and $\mathbf{c}$ are collinear.

Remark 5: It makes sense that two points, say **1** and **2**, coincide. Then the chord degenerates to the tangent at $\mathbf{1} = \mathbf{2}$. Corresponding configurations where one, two, or three pairs of points coincide are shown in Figure 22.10.

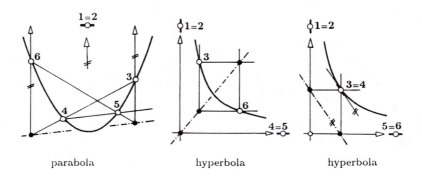

parabola   hyperbola   hyperbola

Figure 22.10: Degenerate hexagons.

Remark 6: Pascal's theorem can be used to construct additional points or tangents. An example is given in Figure 22.11.

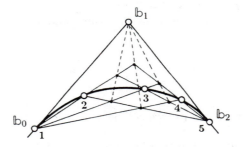

Figure 22.11: Bézier points of a five-point conic.

22.6 Brianchon's Theorem

Translating Pascal's theorem into its dual version gives Brianchon's theorem:

> If the sides of a hexagon are tangential to a non-degenerate conic section, then the diagonals intersect in a common point.

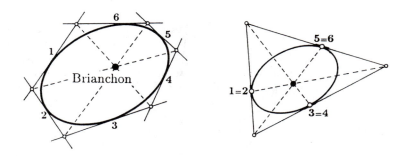

Figure 22.12: Brianchon's Theorems.

If two tangents coalesce, their intersection point becomes the point of contact. Of particular interest is the case of three points with tangents, as illustrated in Figure 22.12. It is due to Ceva (1678).

Remark 7: A nice generalization and proof of Brianchon's theorem due to Chasles (1865) uses no algebra, but some imagination. Consider three conic sections on a quadric. Their planes intersect pairwise in one of three lines, which all meet in one point, while their projections are inscribed on the silhouette of the quadric as shown in Figure 22.13. If the three planes are tangent planes, the three conic sections degenerate to three pairs of lines, while the projection degenerates to Brianchon's configuration. Note that these tangents are real only if the quadric is a hyperboloid. The dual figure proves Pascal's theorem.

Example 1: Brianchon's configuration is also contained in Figure 21.4. The six solid lines are tangents of a conic section. Moreover, if λ varies, the line $\mathcal{L}$ envelopes this conic section. This property is dual to the property discussed in Remark 3.

Figure 22.13: A generalization of Brianchon's Theorem.

22.7 Rational Bézier Curves

A rational Bézier curve is defined by

$$\mathsf{x}(t) = \sum_{i=0}^{n} \mathsf{b}_i B_i^n(t) \;,$$

where $\mathsf{b}_i = [\beta_i \; \mathbf{b}_i^t]^t$ are the so-called **control points** and

$$B_i^n(t) = \binom{n}{i}(1-t)^{n-i} t^i$$

are the Bernstein polynomials of degree n which were already introduced in Section 12.3. Inhomogeneizing gives

$$\frac{\mathbf{x}}{x_0} = \frac{\sum \mathbf{b}_i B_i^n(t)}{\sum \beta_i B_i^n(t)} \;.$$

A projective change of the parameter by

$$t = \frac{s}{1+s}$$

transforms the representation of the curve to **monomial form**,

$$\varkappa\xi = \sum_{i=0}^{n} \mathbb{b}_i \binom{n}{i} s^i \ ,$$

where $\xi = (1+s)^n$. The subsequent dilatation $s = \sigma r$ of the parameter gives

$$\varkappa\xi = \sum \mathbb{b}_i \sigma^i \binom{n}{i} r^i,$$

where the $\mathbb{b}_i \sigma^i$ represent "weighted" control points. Finally, inhomogeneizing gives

$$\frac{\mathbf{x}}{x_0} = \frac{\sum \mathbf{b}_i \sigma^i \binom{n}{i} s^i}{\sum \beta_i \sigma^i \binom{n}{i} s^i} \ .$$

Replacing σ by σ/ϱ, one gets that the curve does not change if the control points $\mathbb{b}_0, \dots, \mathbb{b}_n$ are multiplied by factors $\sigma^0 \varrho^n, \dots, \sigma^n \varrho^0$, respectively. One can apply this fact to put suitable factors to $\mathbb{b}_0$ and $\mathbb{b}_n$.

Remark 8: The steps of de Casteljau's algorithm explained in Section 12.3 are closely related to Brianchon's theorem. Let $\mathbf{p}_i^r$ be the result if one inhomogeneizes $\mathbb{b}_i^r = \mathbb{b}_i^{r-1}(1-t) + \mathbb{b}_{i+1}^{r-1}t$. In order to investigate the construction of the $\mathbf{p}_i^r$ it suffices to consider the A-frame shown in Figure 22.14. As in Example 1, the lines $\mathbf{p}_0^1 \mathbf{p}_1^1$ envelope a conic section for varying t with point of contact $\mathbf{p}_0^2$. Hence, the point of contact can also be constructed by Brianchon's theorem as demonstrated in Figure 22.14. This construction holds for all $\mathbf{p}_i^r, r > 1$, and was already discussed by Haase in 1870.

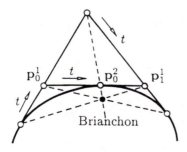

Figure 22.14: Haase's use of Brianchon's Theorem.

22.8 Rational Bézier surfaces

A rational triangular Bézier surface is defined by

$$\varkappa(u, v, w) = \sum_{\triangle} \mathbb{b}_{ijk} B_{ijk}^n(u, v, w) \ ,$$

where $\sum_{\triangle}$ means a sum over all $i, j, k \geq 0$ such that $i + j + k = n$. The
points $\mathbb{b}_{ijk} = [\beta_{ijk} \quad \mathbb{b}_{ijk}^t]^t$ are the so-called **control points**;

$$B_{ijk}^n = \frac{n!}{i!j!k!} u^i v^j w^k, \quad u + v + w = 1;$$

are the **trivariate Bernstein polynomials** of degree n; and u, v, w are
barycentric coordinates in the uv-plane. Inhomogeneizing gives

$$\frac{\mathbf{x}}{x_0} = \frac{\sum \mathbf{b}_{ijk} B_{ijk}^n(u, v, w)}{\sum \beta_{ijk} B_{ijk}^n(u, v, w)} \ .$$

The surface does not change if one drops the stipulation $u+v+w = 1$, which
means that $\mathbb{u} = [u \quad v \quad w]^t$ is allowed to be a homogeneous coordinate
column. Thus one can write

$$\varkappa(\mathbb{u}) = \sum_{\triangle} \mathbb{b}_{ijk} B_{ijk}^n(\mathbb{u}) \ ,$$

where $\mathbb{u}$ are projective coordinates in the uv-plane. The simple transforma-
tion $[u \quad v \quad w] = [\varrho r \quad \sigma s \quad \tau t]$ leaves the reference triangle unchanged, but
it changes the unit point and transforms the representation of the surface
into

$$\varkappa = \sum_{\triangle} \mathbb{b}_{ijk} \varrho^i \sigma^j \tau^k B_{ijk}^n(\mathbb{r}) \ ,$$

where the $\mathbb{b}_{ijk} \varrho^i \sigma^j \tau^k$ represent "weighted" control points. Inhomogeneizing gives

$$\frac{\mathbf{x}}{x_0} = \frac{\sum \mathbf{b}_{ijk} \varrho^i \sigma^j \tau^k B_{ijk}^n(\mathbb{r})}{\sum \beta_{ijk} \varrho^i \sigma^j \tau^k B_{ijk}^n(\mathbb{r})} \ .$$

One can use this reparametrization to put suitable factors to $\mathbb{b}_{n00}$, $\mathbb{b}_{0n0}$,
and $\mathbb{b}_{00n}$. Stipulating $r + s + z = 1$ guarantees uniqueness of the representation.

Remark 9: Inhomogeneizing the parameter u by $s = u/w$ and $t = v/w$ gives the bivariate monomial form

$$\varkappa\xi = \sum_{\triangle} \mathrm{b}_{ijk}\frac{n!}{i!j!k!}s^i t^j,$$

where $\xi = 1/w^n$.

22.9 Notes and Problems

1 In Section 23.1 it is shown that certain points of a complete quadrangle lie in harmonic position. This fact can also be proven analytically with the four vertices • of the quadrangle forming the coordinate frame.

2 Given three points on a straight line, a complete quadrangle can be used to construct the fourth harmonic point.

3 Desargues' configuration in Figure 22.4 can be interpreted as the projection of the intersections of two planes with a pyramid whose apex is projected onto z.

4 Any two figures in perspective position with respect to z can be viewed as the projection of the intersections of two hyperplanes with a general cone whose apex is projected onto z. An example is given in Figure 22.5.

5 Desargues' theorem can be proven by successive applications of Pappus' theorem (Hessenberg 1905).

6 Desargues' configuration contains 10 pairs of triangles in perspective position.

7 Pappus' theorem is closely related to the **commutative law** of multiplication (Hilbert 1899), as illustrated by the lettering of the affine scales in Figure 22.6.

8 Pascal's theorem is closely related to Steiner's generation of conics, which is mentioned in Remark 4. The dual relationship is discussed in Example 1.

9 Consider a hexagon with parallel opposite sides and 5 vertices on a circle. Then opposite angles are equal, and the two chords **24** and **15** have equal lengths for symmetry reasons. Thus, the theorem on angles of circumference implies that the sixth point also lies on the circle. See Section 17.3.

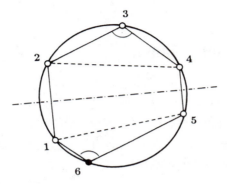

Figure 22.15: Proof of Pascal's theorem.

23 Projective Quadrics

The projective viewpoint is the most general. This is of particular significance for quadrics. Most of their interesting and useful affine properties stem from only a few projective properties and from concepts such as polarity.

Literature: Berger, Blaschke, Samuel, Wylie

23.1 Projective Quadrics

A **projective quadric** in $\mathcal{P}^n$ consists of all points in $\mathcal{P}^n$ satisfying a quadric equation. Hence the equation of a projective quadric can be written as

$$Q(\varkappa) = \varkappa^t \mathbb{C} \varkappa = 0 \ ,$$

where $\mathbb{C}$ is a non-zero symmetric $n+1 \times n+1$ matrix. As before, the letter Q is also used to denote the quadric itself. Writing

$$\mathbb{C} = \begin{bmatrix} c & \mathbf{c}^t \\ \mathbf{c} & C \end{bmatrix} \quad \text{and} \quad \varkappa = \begin{bmatrix} x_0 \\ \mathbf{x} \end{bmatrix} \ ,$$

one has

$$Q(\varkappa) = \mathbf{x}^t C \mathbf{x} + 2\mathbf{c}^t \mathbf{x} x_0 + c x_0^2 = 0 \ .$$

The plane $x_0 = 0$ is regarded as the ideal plane. Then Q has an affine and an ideal component, Q_a and Q_∞. Their equations are obtained from

$Q(\varkappa) = 0$ by setting $x_0 = 1$ and $x_0 = 0$, respectively. Obviously, Q_a is an **affine quadric** while Q_∞ is a projective quadric which consists of the asymptotic directions of Q_a. The quadric Q is called the **projective extension** of Q_a, while Q_∞ is referred to as the **ideal conic section** or **quadric** corresponding to Q_a.

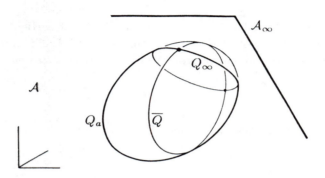

Figure 23.1: Projective extension of a quadric.

Remark 1: If Q_∞ is a real conic section, Q is called a **hyperboloid**. If Q_∞ is non-real, Q is called an **ellipsoid**. Finally, if the ideal plane is tangent to Q, Q is called a **paraboloid**.

23.2 Tangent Planes

This section extends the material on affine quadrics from Section 13.4 to projective quadrics. Consider a line T in $\mathcal{P}^n$ given by $\varkappa = \mathfrak{q} + \lambda \mathfrak{p}$, where λ is a projective scale on T. Let $\mathfrak{q}$ be a point of the quadric Q. Then T is tangent to Q at $\mathfrak{q}$ if $\mathfrak{q}$ is the only intersection of T with Q or if T lies completely on Q. Thus, T is tangent to Q if the quadratic equation in λ,

$$Q(\varkappa) = \varkappa^t \mathbb{C} \varkappa = \lambda^2 \mathfrak{p}^t \mathbb{C} \mathfrak{p} + 2\lambda \mathfrak{q}^t \mathbb{C} \mathfrak{p} = 0 ,$$

has a double root at $\lambda = 0$. Hence T is tangent to Q if $\mathfrak{q}^t \mathbb{C} \mathfrak{p} = 0$. Consequently,

$$\mathfrak{u}^t \varkappa = \mathfrak{q}^t \mathbb{C} \varkappa = 0$$

describes the **tangent hyperplane** of Q at q.

The tangent plane is not defined if $\mathsf{C}^t\mathsf{q} = \mathsf{o}$. In this case q is called a **singular point** of Q, and Q is called **degenerate**. The set of singular points of Q forms a projective subspace.

A given u represents a tangent plane of Q at a point q if the homogeneous system

$$\mathsf{Cq} - \mathsf{u}\varrho = \mathsf{o}$$
$$\mathsf{u}^t\mathsf{q} \quad\ = 0$$

is solvable for q and ϱ, i.e., if

$$\det \begin{bmatrix} \mathsf{C} & \mathsf{u} \\ \mathsf{u}^t & 0 \end{bmatrix} = 0 \ .$$

The above equation is a quadratic equation in the u_i called the **tangential equation** of Q.

The **polar plane** of some arbitrary point q in $\mathcal{P}^n$ with respect to Q is defined by

$$\mathsf{q}^t\mathsf{Cx} = 0 \ .$$

Obviously, all tangent planes of Q through q contact Q in points of the polar plane, and the polar plane is a tangent plane of Q in the case where q lies on Q. The polar plane is not defined if q is a singular point. Since $\mathsf{q}^t\mathsf{Cx} = 0$ is symmetric in q and x, one also has that q lies in the polar plane of x. Such pairs of points are called **conjugate with respect to** Q. With this terminology, the points of Q are **self-conjugate**.

Remark 2: If C is non-singular, the **polarity** $\mathsf{u} = \mathsf{Cx}$ represents a correlation. The image u is a tangent plane of Q if it contains x or equivalently if x is a point on Q since

$$\mathsf{u}^t\mathsf{x} = \mathsf{x}^t\mathsf{C}^t\mathsf{x} \ .$$

Recall that C is symmetric and invertible. Therefore, since $\mathsf{x} = \mathsf{C}^{-1}\mathsf{u}$, the tangential equation above can be written as

$$\mathsf{u}^t\mathsf{C}^{-1}\mathsf{u} = 0 \ .$$

Remark 3: More generally, the planes u satisfying a quadratic equation

$$\mathsf{u}^t \mathbb{D} \mathsf{u} = 0 \;, \quad \mathbb{D}^t = \mathbb{D} \neq \mathbb{O}$$

form a **tangential quadric,** as opposed to the **point quadric** formed by the points x satisfying the **point equation** $\mathsf{x}^t \mathbb{C} \mathsf{x} = 0$. Obviously, the tangent planes of a non-degenerate point quadric form a tangential quadric. But a pair of planes is not a tangential quadric, whereas the dual configuration, a pair of points, is a tangential quadric. Dual to a cone is a conic section enveloped by its tangent planes. Examples are given in Figures 20.5 and 20.6.

Example 1: The tangent planes of the ellipsoid

$$\frac{x_1^2}{\lambda_1} + \frac{x_2^2}{\lambda_2} + \frac{x_3^2}{\lambda_3} = x_0^2$$

satisfy the tangential equation

$$\lambda_1 u_1^2 + \lambda_2 u_2^2 + \lambda_3 u_3^2 = u_0^2 \;.$$

Example 2: For $\lambda_3 \to 0$ the ellipsoid converges to the ellipse

$$\frac{x_1^2}{\lambda_1} + \frac{x_2^2}{\lambda_2} = x_0^2 \;, \qquad x_3^2 = 0 \;,$$

whose tangent planes satisfy the tangential equation

$$\lambda_1 u_1^2 + \lambda_2 u_2^2 = u_0^2 \;.$$

23.3 The Role of the Ideal Plane

Let p be an ideal point, then $p_0 = 0$ and the polar plane of p is given by

$$\mathbf{p}^t \left[C\mathbf{x} + \mathbf{c} x_0 \right] = 0 \;.$$

This equation also represents the diametric plane $\mathcal{V}$ of Q_a with respect to the direction $\mathbf{v} = \mathbf{p}$. This means that a diametric plane can simply be viewed as the polar plane of an ideal pole, as illustrated in Figure 23.2.

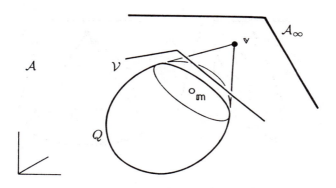

Figure 23.2: Diametric plane.

A diametric plane contains all midpoints of Q_a. Since the relationship between a pole and the points of its polar plane is symmetric, every midpoint of Q_a is a point whose polar plane is the ideal plane. Consequently, each tangent plane of Q through a midpoint **m** of Q_a contacts Q at an ideal point v. This explains why the cone from $\mathcal{M}$ to Q_∞ is called the **asymptotic cone**.

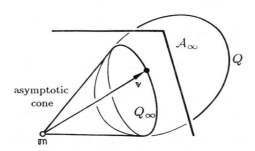

Figure 23.3: Midpoint and asymptotic cone.

Furthermore, an axial direction corresponds to a singular point of Q_∞, while a singular axial direction corresponds to a singular point of Q_∞ which is also a singular point of Q.

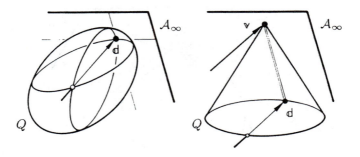

Figure 23.4: Axial directions .

23.4 Harmonic Points and Polarity

Polarity and harmonic position are closely related. Let $\mathbb{p}$ and $\mathbb{q}$ be points conjugate with respect to Q, i.e., $\mathbb{p}^t C \mathbb{q} = 0$, and suppose $\mathbb{p}$ is not on Q. Then, intersecting the line $\varkappa = \mathbb{p} + \lambda \mathbb{q}$ with Q gives

$$Q(\varkappa) = \mathbb{p}^t C \mathbb{p} + \lambda^2 \mathbb{q}^t C \mathbb{q} = \varrho(\lambda - \lambda_0)(\lambda + \lambda_0) = 0 .$$

Thus the intersection points are

$$\varkappa_+ = \mathbb{p} + \lambda_0 \mathbb{q} \qquad \text{and} \qquad \varkappa_- = \mathbb{p} - \lambda_0 \mathbb{q} .$$

Since $cr[-\lambda_0 + \lambda_0 | 0 \infty] = -1$, the intersection points lie in harmonic position with respect to $\mathbb{p}$ and $\mathbb{q}$. This property characterizes the polarity.

Note that $\varkappa_+$ and $\varkappa_-$ may be a complex conjugate pair or may coincide (with $\mathbb{q}$).

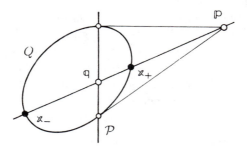

Figure 23.5: Polarity and harmonic position.

23.5 Pencils of Quadrics

Let Q_0 and Q_1 be two quadrics given by the equations $\varkappa^t \mathbb{C}_0 \varkappa = 0$ and $\varkappa^t \mathbb{C}_1 \varkappa = 0$, respectively. Then the family of quadrics defined by

$$Q(\varkappa) = \varkappa^t \left[\lambda_0 \mathbb{C}_0 + \lambda_1 \mathbb{C}_1 \right] \varkappa = 0 \ ,$$

more concisely written as $Q = Q_0 + \lambda Q_1$, is said to form a **pencil** spanned by the quadrics $\mathbb{Q}_0$ and $\mathbb{Q}_1$ and controlled by the homogeneous coordinates λ_0, λ_1 of the parameter λ. It has the following simple properties:

The pencil is spanned by any two of its members, and all of its members contain the intersection C of Q_0 and Q_1.

Every point not in the common intersection lies on exactly one member of the pencil.

The family of intersections with any subspace (in general position) forms a pencil again.

If $\det \left[\lambda_0 \mathbb{C}_0 + \lambda_1 \mathbb{C}_1 \right] = 0$, then Q contains a singular point, i.e., Q is a cone.

In $\mathcal{P}^n$, there are, in general, n members of the pencil tangential to a given hyperplane since, in general, the corresponding tangential equation is of degree n.

Example 3: In the case where Q_1 is a pair of planes in $\mathcal{P}^3$ and Q_0 is non-degenerate, the quadric Q_1 intersects Q_0 in two conic sections, forming the intersection C, as illustrated in Figure 23.6. All members of the pencil contain these conic sections, but the conic sections themselves cannot belong to the pencil. If Q_0 and Q_1 are in general position, this pencil contains two cones. However, if the pair of planes Q_1 is in some special position to Q_0, the cones degenerate.

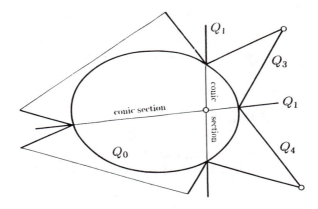

Figure 23.6: Pencil containing a pair of planes.

Example 4: The case where Q_1 is a double plane is discussed in Chapter 15, where Q_1 is given by $x_0^2 = 0$. Let Q_0 be non-degenerate and such that $x_0 = 0$ is not a tangent plane of Q_0. Then one can conclude from Chapter 15 and Section 23.3 that all members of the pencil contact each other at the intersection C of Q_0 and Q_1 and that there is exactly one proper cone in the pencil. This is illustrated in Figure 23.7. The pencil degenerates if the double plane is tangential to Q_0. This case is also discussed in Chapter 15.

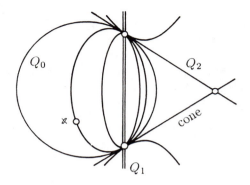

Figure 23.7: Pencil containing a double plane.

23.6 Ranges of Quadrics

Dual to a pencil one can consider pencils of tangential quadrics in the dual space given by

$$Q^*(\mathsf{u}) = \mathsf{u}^t \left[\mu_0 \mathbb{D}_0 + \mu_1 \mathbb{D}_1 \right] \mathsf{u} = 0 \ .$$

Such a pencil is called a **range** or **tangential pencil**. Its simple properties are dual to those of a pencil. All its members contain the common tangent planes of Q_0^* and Q_1^*. There is exactly one quadric in a range containing a given plane, but in general n quadrics meet at a given point. The number of quadrics meeting at a point is less than n if Q_0^* and Q_1^* are in special positions to each other. Consequently, the tangential quadrics corresponding to the quadrics of a pencil can lie in a range only if this number equals one. This is the case only if the pencil contains a double plane or if the range contains a double point. See Examples 4 and 6.

Example 5: The range of quadrics dual to the pencil in Example 3 contains a pair of points, say Q_1^*. If Q_0^* is non-degenerate, the common tangent planes of Q_0^* and Q_1^* form a pair of cones. All members of the range are inscribed in these two cones which themselves cannot belong to the range. If Q_0^* and Q_1^* are in general position the range contains two conic sections given by the intersection of both cones. The range further degenerates if the pair of points Q_1^* is in some special position to Q_0^*.

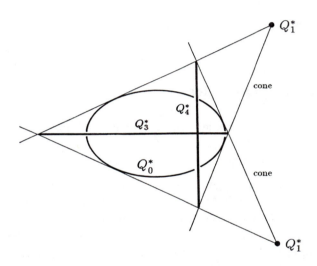

Figure 23.8: Range containing a pair of points.

Example 6: The range dual to the pencil of Example 4 contains a double point Q_1^* and a conic section Q_2^* which are the only degenerate members of the range, cf. Remark 3. Moreover, the point equation of the non-degenerate quadrics of this range is linear in μ. Hence, the corresponding point quadrics lie in a pencil, as in Example 4.

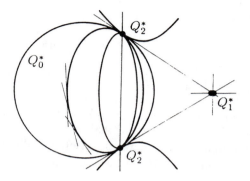

Figure 23.9: Range containing a double point.

23.7 The Imaginary in Projective Geometry

A projective line which does not lie on a quadric intersects this quadric in exactly two points, if one includes and counts ideal points, complex conjugate points, and also multiple points. This extension simplifies some of the discussions above and allows a deeper insight into the main features of quadrics.

The same is true when one permits imaginary points for planes, quadrics, and projective configurations in general. This extension is performed in a very straightforward way, by regarding imaginary points, lines, planes, etc. as if they were real ones. Even this naive point of view gives amazing insight into common properties of certain projective configurations, which seem very different at first glance.

In $\mathcal{P}^2$, for example, the "hyperbola" $x_1 x_2 = x_0^2$ can be mapped onto the "circle" $y_1^2 + y_2^2 = y_0^2$ by the imaginary affine map

$$ y = \begin{bmatrix} 2 & 0 & 0 \\ 0 & 1 & 1 \\ 0 & -i & i \end{bmatrix} x \ . $$

Its asymptotic cone $x_1 x_2 = 0$ is mapped onto the asymptotic cone $y_1^2 + y_2^2 = 0$ which intersects the ideal line at

$$
\mathring{\imath}_+ = \begin{bmatrix} 0 \\ 1 \\ i \end{bmatrix} \qquad \text{and} \qquad \mathring{\imath}_- = \begin{bmatrix} 0 \\ 1 \\ -i \end{bmatrix} .
$$

The directions $\mathbf{i}_- = [1 \ {-i}]^t$ and $\mathbf{i}_+ = [1 \ \ i]^t$ have incredible properties in the Euclidean plane $\mathcal{E}^2$. Each is orthogonal to itself and has zero length. Hence, these directions are called **minimal** or **isotropic directions** and lines in these directions are called **minimal** or **isotropic lines**. Note that any two points on such a line have distance 0.

Moreover, every circle in the projective extension of $\mathcal{E}^2$ contains the so-called **circular points** $\mathring{\imath}_-$ and $\mathring{\imath}_+$, and any conic section containing the circular points is a circle. The two lines from its midpoint $\mathfrak{m}$ to the circular points are the asymptotes of all concentric circles, as illustrated in Figure 23.10, where imaginary elements are dashed and the ideal line is represented by a real line.

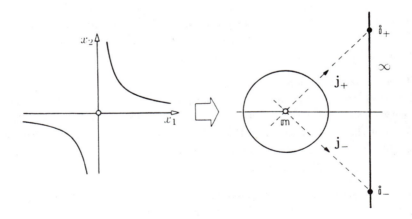

Figure 23.10: The circular points.

In $\mathcal{P}^3$, for example, the "hyperbolic paraboloid" $x_1 x_2 = x_3 x_0$ can be mapped onto the "sphere" $y_1^2 + y_2^2 + y_3^2 = y_0^2$ by the imaginary collineation

$$y = \begin{bmatrix} 1 & 0 & 0 & 1 \\ 0 & 1 & 1 & 0 \\ 0 & -i & i & 0 \\ -1 & 0 & 0 & 1 \end{bmatrix} x \ .$$

This collineation maps the net of generatrices, which may be written as

$$x = \begin{bmatrix} 1 & \lambda & \mu & \lambda\mu \end{bmatrix},$$

onto the net described by

$$y = \begin{bmatrix} \lambda\mu + 1 \\ \mu + \lambda \\ i(\mu - \lambda) \\ \lambda\mu - 1 \end{bmatrix} \ .$$

This net represents the two non-real families of straight lines $\lambda = $ *fixed* and $\mu = $ *fixed* on a sphere. For $\mu\lambda = -1$ the points y of the net lie on the ideal plane where they form the conic section

$$y_1^2 + y_2^2 + y_3^2 = 0, \quad y_0 = 0 \ .$$

This conic section is called the **spherical circle**, and it has amazing properties in $\mathcal{E}^3$ similar to those of the circular points in $\mathcal{E}^2$. In particular, any direction corresponding to one of its points is minimal and any sphere in the projective extension of $\mathcal{E}^3$ contains this imaginary circle. This is a characteristic property of spheres. Moreover, the spherical circle intersects any plane in its circular points. Thus each plane intersection of a sphere is a circle.

Remark 4: The point y above is real if and only if λ and μ are complex conjugate. By introducing new parameters $\lambda = z_1 + iz_2$, $\mu = z_1 - iz_2$ one obtains for the sphere the parametrization

$$y = \begin{bmatrix} z_1^2 + z_2^2 + 1 \\ 2z_1 \\ 2z_2 \\ z_1^2 + z_2^2 - 1 \end{bmatrix}$$

which represents the so-called **stereographic projection**, where $z = [z_1 \quad z_2]^t$ are the Cartesian coordinates of a point in the plane $z_3 = 0$, as illustrated in Figure 23.11. Note that the minimal lines of this plane correspond to the imaginary generatrices of the sphere.

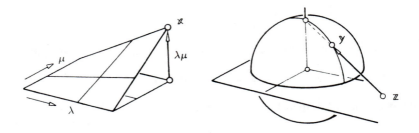

Figure 23.11: Stereographic projection.

Remark 5: According to the real and imaginary parts of a projective scale, there are ∞^1 real points on a projective line but ∞^2 imaginary points since the imaginary points depend on two real parameters.

As a consequence, given two real points x and y of a line, which are separated by two real points $\mathfrak{a}$ and $\mathfrak{b}$, it is possible to walk from x to y without meeting $\mathfrak{a}$ or $\mathfrak{b}$ by going through the complex domain. Also recall Note 7 of Section 21.7.

In the same way, one finds ∞^4 imaginary points in a projective plane and ∞^6 imaginary points in a projective space. In a pencil or bundle one has ∞^2 and ∞^4 imaginary members, respectively.

23.8 The Steiner Surface

Every plane intersects the parabolic Steiner surface of Section 16.8 at real points. Hence there exists no real projective map which maps the surface into a finite surface. However, the imaginary map

$$x_1 = iy_0 \; , x_2 = y_2 + y_3 \; , x_3 = -y_2 + y_3 \; , cx_0 = iy_1 \; ,$$

transforms this parabolic surface into the surface

$$y_2^2 y_3^2 + y_3^2 y_1^2 + y_1^2 y_2^2 = 2 y_1 y_2 y_3 y_0$$

which can be viewed as the projective **normal form** of **Steiner's surface**. It is easy to check by substitution that this surface is parametrically given by

$$y_1 = 2vw \ , y_2 = 2wu \ , y_3 = 2uv \ , y_0 = u^2 + v^2 + w^2 \ .$$

Figure 23.12 shows this finite surface. It is inscribed in the tetrahedron formed by the four planes

$$-x + y + z = 1 \ , x - y + z = 1 \ , x + y - z = 1 \ , -x - y - z = 1 \ .$$

It touches these four planes in four conic sections which are inscribed in the triangular sides. These four conic sections touch one another at the midpoints of the 6 edges. These points are called the **cuspidal points** of the surface. There are three pairs of opposite cuspidal points. The three lines connecting these pairs are double lines of the surfaces where it intersects itself. These three lines meet at a threefold point of the surface.

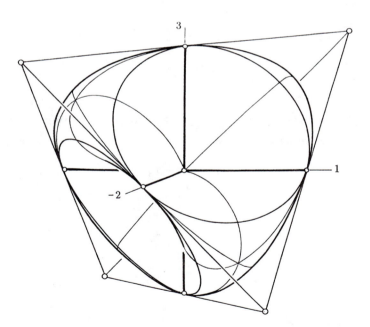

Figure 23.12 : Steiner's surface, finite example.

Remark 6: Any non-degenerate Steiner surface can be transformed into the normal form above. On the other hand any triangular rational quadratic patch defines a degenerate or non-degenerate Steiner surface. In particular the quadratic patch defines a quadric if the three patch boundaries meet in one point and have coplanar tangents there. This condition is necessary for the triangular patch to lie on a non-degenerate quadric. However, a triangular rational quadratic patch can lie on a cone without satisfying this condition.

23.9 Notes and Problems

1 A quadric on a line is either a pair of real, conjugate complex, or coalescing points.

2 A conic section can be transformed by a real collineation into one of the 5 **projective normal forms**:

$$x_0^2 + x_1^2 + x_2^2 = 0$$
$$x_0^2 + x_1^2 - x_2^2 = 0$$
$$x_1^2 + x_2^2 = 0$$
$$x_1^2 - x_2^2 = 0$$
$$x_2^2 = 0 \ .$$

3 A quadric in $\mathcal{P}^3$ can be transformed by a real collineation into one of the 8 **projective normal forms**:

$$x_0^2 + x_1^2 + x_2^2 + x_3^2 = 0$$
$$x_0^2 + x_1^2 + x_2^2 - x_3^2 = 0$$
$$x_0^2 + x_1^2 - x_2^2 - x_3^2 = 0$$
$$x_1^2 + x_2^2 + x_3^2 = 0$$
$$x_1^2 + x_2^2 - x_3^2 = 0$$
$$x_2^2 + x_3^2 = 0$$
$$x_2^2 - x_3^2 = 0$$
$$x_3^2 = 0 \ .$$

4 In general, real quadrics can be distinguished by the dimensions of their real generators and their singular subspaces.

5 If one admits imaginary collineations, quadrics differ only in the dimensions of their singular subspaces.

6 Any subspace intersects a pencil of quadrics in a pencil of quadrics. In particular, any line in general position intersects a pencil of quadrics in points of a so-called **involution** (Desargues 1639).

7 The intersection of two planes given by $\mathsf{u}^t\mathsf{x} = 0$ and $\mathsf{v}^t\mathsf{x} = 0$ is tangent to a non-degenerate quadric $\mathsf{x}^t\mathbb{C}\mathsf{x} = 0$ if

$$\det \begin{bmatrix} \mathbb{C} & \mathsf{u} & \mathsf{v} \\ \mathsf{u}^t & 0 & 0 \\ \mathsf{v}^t & 0 & 0 \end{bmatrix} = 0 .$$

8 Circular points and the spherical circles are transformed onto themselves under Euclidean motions and centric dilatations.

9 The tangential equations of the circular points and of the spherical circle are

$$u_1^2 + u_2^2 = 0 \qquad \text{and} \qquad u_1^2 + u_2^2 + u_3^2 = 0 ,$$

respectively.

10 Confocal quadrics in $\mathcal{E}^2$ and $\mathcal{E}^3$ are defined as members of a range which contains the circular points or the spherical circle, respectively.

11 There is a line of real points in every imaginary plane. If the plane is given by $[\mathsf{u} + \mathsf{v}i]^t\mathsf{x} = 0$, the real line is the intersection of $\mathsf{u}^t\mathsf{x} = 0$ and $\mathsf{v}^t\mathsf{x} = 0$.

12 In $\mathcal{E}^2$ the angle φ between two directions **a** and **b** can be computed by

$$\varphi = \frac{1}{2} \ln cr \, [\mathbf{ab}|\mathbf{i}_+\mathbf{i}_-] .$$

13 With this definition the property discussed in Section 22.4 leads to the constant angle of circumference in Section 17.3.

14 Consider a quadric Q in $\mathcal{P}^2$ and a line $\mathcal{X}$. The points conjugate to all points of $\mathcal{X}$ form a line $\mathcal{Y}$, and $\mathcal{X}$ and $\mathcal{Y}$ are called **conjugate polars** with respect to Q.

15 Each surface which contains ∞^2 conic sections is either a quadric or a cubic ruled surface or a Steiner surface (Darboux 1880).

16 Each surface whose plane intersections are rational is a ruled rational surface or a Steiner surface (Picard 1878).

PART SIX

Some
Descriptive Geometry

Descriptive Geometry is concerned with the presentation of spatial objects in one or two planar projections and with the rules of generating such representations. It forms the theoretical basis of architectural and all technical drafting. Usually, an object is projected orthogonally into a pair of orthogonal planes which are in natural positions relative to the object. Such projections lend themselves to an easy reconstruction of the spatial object, and they allow one to draw any other projection of it.

By the methods of Descriptive Geometry, one also can construct the intersection of two surfaces as well as special curves and surfaces, such as a helix or a quadric. The methods of Descriptive Geometry go back more than two millenia, but it was the French mathematician Caspard Monge (1746-1818) who first systematized the many methods in use and who reduced them to a small number of simple principles.

24 Associated Projections

After projecting a spatial object into a plane, one dimension is lost. This loss can be compensated for by integrating, for example, the ground plan into the projection or by entering the heights into the plan, as illustrated in Figure 24.1. A more common approach, however, is to draw two orthogonal projections. Two such projections of a three-dimensional object are interrelated and facilitate the solutions to most kinds of construction problems.

Literature: Hohenberg, Koenderink, Leighton, Wunderlich

24.1 Plan and Elevation

Consider an object in space given by the Cartesian coordinates of its points $\mathbf{x} = [x \quad y \quad z]^t$. The points $\mathbf{x}' = [x \quad y]^t$ represent its orthogonal projection into the plane $z = 0$. This projection is called **plan**. Similarly, the points $\mathbf{x}'' = [y \quad z]^t$ represent the orthogonal projection into the plane $x = 0$ which is called **elevation**. In order to facilitate comparisons, the plan and the elevation are drawn with a common y-axis. Two such projections are called **associated**.

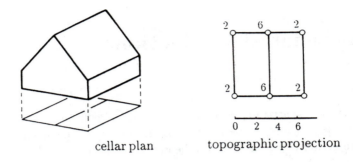

cellar plan   topographic projection

Figure 24.1: Cellar plan and topographic projection.

Visibility can be decided immediately with the help of an associated projection. If two points $\mathbf{x}_1$ and $\mathbf{x}_2$ have the same plan $\mathbf{x}'_1 = \mathbf{x}'_2$ and $z_1 > z_2$, then $\mathbf{x}'_1$ hides $\mathbf{x}'_2$.

The common y-axis denoted by $*$ in the figures is called the **ground line**. The line connecting $\mathbf{x}'$ and $\mathbf{x}''$ is called the **order line**. It is perpendicular to the ground line because of the common y-coordinate.

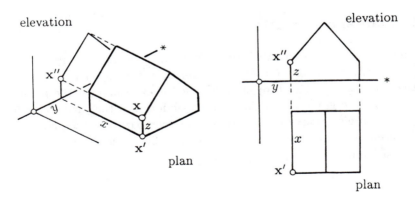

Figure 24.2: Plan and elevation.

Remark 1: A point on a line or a surface is already determined by the line or the surface and one of its projections. Figure 24.3 shows three examples of how to find the associated projection.

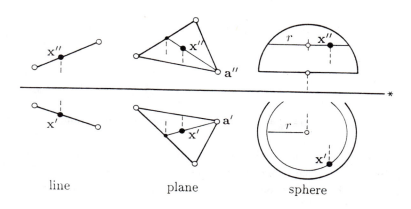

Figure 24.3: United positions.

24.2 Side Elevation

Usually the Cartesian reference system is closely related to the edges and planes of the object, as illustrated in Figure 24.2. Sometimes, however, a change of the vertical plane is desired; this corresponds to a Euclidean motion of the xy-plane in itself. Constructing a new elevation is simple; the rules are illustrated in Figure 24.4. The new order lines are perpendicular to the new ground line marked by $**$ while the z-coordinates remain the same. An example is shown in Figure 24.5. The new projection is called a **side elevation**; it can replace the elevation.

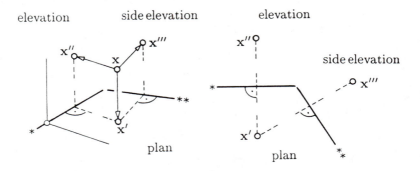

Figure 24.4: Principle of drawing a side elevation.

Notice that one can use the same procedure to change the plan view while maintaining the same elevation plane. Then the new ground line lies in the elevation plane while the x-coordinates remain the same.

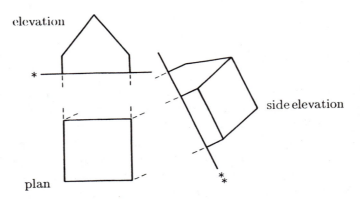

Figure 24.5: General side elevation, example.

24.3 Special Side Elevations

Special side elevations can simplify the reconstruction of spatial configurations. In particular, the following special positions are shown in Figure 24.6:

(a) An elevation parallel to some line shows the true length of this line.
(b) An elevation orthogonal to a horizontal line shows this line as a point.
(c) An side elevation orthogonal to a plane shows this plane as a line.
(d) An elevation parallel to a plane figure shows its true shape.

The use and the construction of special side elevations is a simple and effective tool to solve many problems in Descriptive Geometry.

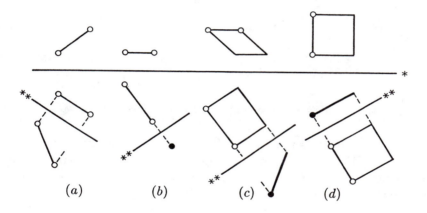

Figure 24.6: Special side elevations.

Example 1: The shadow of a telegraph-post on a farm house can be constructed with the aid of rule (c). From the side elevation where the roof degenerates into a line segment, one can obtain the intersection of the light rays with the roof.

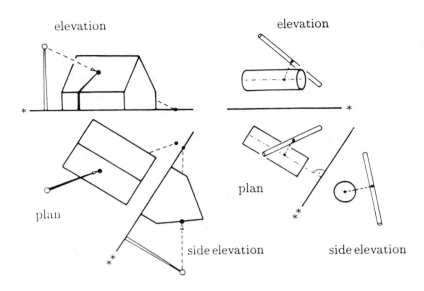

elevation   elevation

plan   plan

side elevation   side elevation

Figure 24.7: Shadow on a roof and shortest distances between two pipes.

Example 2: Given two pipes as in Figure 24.7, one can detect a possible intersection. Using rule (b), one obtains a side elevation orthogonal to the horizontal pipe which reveals its position in space and the shortest distance to the other pipe.

Example 3: To construct a view of the unit cube along one of its diagonals, one can use rule (a) to generate a side elevation showing the diagonal in its true length. Then a second side elevation orthogonal to the diagonal provides the solution since, according to rule (b), the diagonal is mapped onto one point $\mathbf{a}^{iv}$.

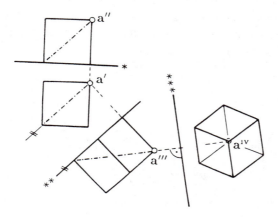

Figure 24.8: Two successive special side elevations of a cube.

24.4 Cross Elevation

The projection onto the xz-plane is called a **cross elevation**. It is a special side elevation which may replace the plan or the elevation, but may also complement both. Figure 24.9 shows two demonstrative examples.

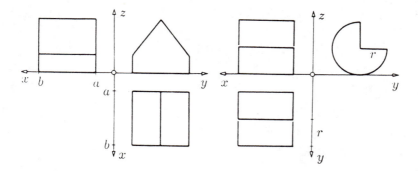

Figure 24.9: Cross elevation, examples.

A cross elevation can also be used to eliminate the parameter t of a planar parametric curve, as demonstrated in Figure 24.10. Note that a minimum (maximum) of $y(x)$ corresponds to a minimum (maximum) of $y(t)$ and vice versa. The same is true for $x(y)$ and $x(t)$.

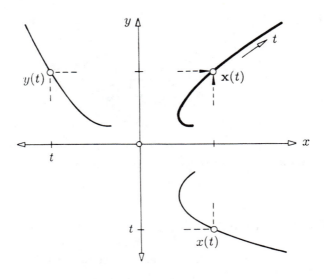

Figure 24.10: Eliminating the parameter.

Moreover, there is a nice relationship between the curvatures $1/R$ and $1/\varrho$ at a minimum of $y(t)$ and the corresponding minimum of $y(x)$. Namely one has

$$\varrho = R \tan^2 \beta,$$

where $\tan\beta = \dot{x}$ is the corresponding slope of $x(t)$, see Figure 24.11. For a proof one only needs to consider a parabola $2R \cdot y = t^2$, $x = t \cdot \tan\beta$. Its cross elevation is a parabola again, $2\varrho \cdot y = x^2$. Hence eliminating t reveals the above relation between ϱ and R.

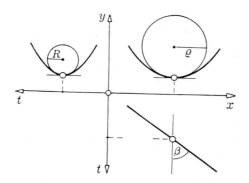

Figure 24.11: Cross elevation and curvature.

Example 4: Cross elevations can also be drawn for control polygons. Consider a Bézier curve $\mathbf{b}(t)$ controlled by the points $\mathbf{p}_i = [x_i \quad y_i]^t$, $i = 0, 1, \ldots, n$. Its coordinates $x(t)$ and $y(t)$ are controlled by the points $[t_i \quad x_i]^t$ and $[t_i \quad y_i]^t$, respectively, where $t_i = i/n$.

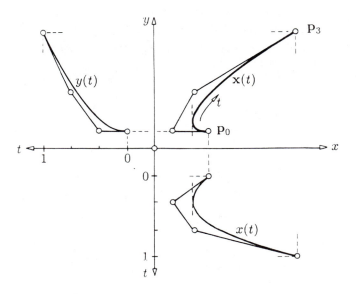

Figure 24.12: Cross elevation of a Bézier curve.

24.5 Curves on Surfaces

All points whose coordinates $\mathbf{x} = \begin{bmatrix} x & y & z \end{bmatrix}^t$ satisfy an equation $F(\mathbf{x}) = 0$ form a spatial surface. Let $\mathbf{x} = \mathbf{x}(t)$ be a curve on this surface, then $c(t) = F(\mathbf{x}(t)) = 0$. Differentiating $c(t)$ with respect to t gives

$$\dot{c} = F_x \dot{x} + F_y \dot{y} + F_z \dot{z} = 0$$

or $\mathbf{n}^t \dot{\mathbf{x}} = 0$, where $\mathbf{n}^t = \begin{bmatrix} F_x & F_y & F_z \end{bmatrix}$ is the so-called **gradient** of F at $\mathbf{x}$. The tangential direction $\dot{\mathbf{x}}$ of the curve at $\mathbf{x}$ is also a tangential direction of the surface at $\mathbf{x}$. Therefore

$$\mathbf{n}^t \begin{bmatrix} \mathbf{y} - \mathbf{x} \end{bmatrix} = 0$$

describes the tangent plane of $F(\mathbf{x}) = 0$ at $\mathbf{x}$. If $\mathbf{n} = \mathbf{o}$, this plane is undefined and $\mathbf{x}$ is called a **singular point** of the surface.

Obviously, if $F_z = 0$ at $\mathbf{x}$, the tangent plane contains the z-direction. Hence, the intersection of the two surfaces $F = 0$ and $F_z = 0$ results in the **contour** of the surface with respect to z. The plan of the contour is the **outline** of the silhouette while the **silhouette** can be interpreted as the shadow of the surface. Figure 24.13 gives an illustration.

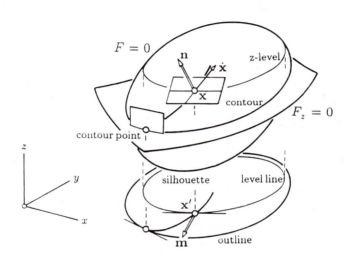

Figure 24.13: Contour, outline, and level line.

At a point where the curve $\mathbf{x}(t)$ crosses the contour, both tangents are projected onto one line, as illustrated in Figure 24.13. In particular, if $\dot{\mathbf{x}}$ is parallel to the z-axis, the projected curve generally has a **cusp**.

A planar curve on the surface given by $F = 0$ and $z = z_0$ is called a **level line** with respect to z. Each level line together with the projection rays to the plan form a cylinder. The tangent planes of these cylinders are given by

$$\mathbf{m}^t\,[\mathbf{y} - \mathbf{x}] = 0 \ ,$$

where $\mathbf{m}^t = [G_x \quad G_y \quad 0]$ denotes the **gradient** of the cylinder $G(x, y, z) = F(x, y, z_0)$. Note that $\mathbf{m}$ agrees with the plan projection of $\mathbf{n}$.

Remark 2: The family of all level lines defines the surface (Cavalieri 1598-1647).

24.6 Canal Surface

A sphere $S = [\mathbf{x} - \mathbf{a}]^t [\mathbf{x} - \mathbf{a}] - r^2 = 0$, whose center $\mathbf{a} = \mathbf{a}(t)$ moves on a curve while its radius $r = r(t)$ varies, envelopes a certain tubular surface, a so-called **canal surface**. The intersection of two spheres $S(t)$ and $S(t + \Delta t)$ lies on their radical plane, $\mathbf{u}^t \mathbf{x} = (S(t + \Delta t) - S(t))/\Delta t = 0$. As Δt becomes infinitesimally small, this plane converges to the plane

$$-\frac{1}{2}\dot{S}(t) = [\mathbf{x} - \mathbf{a}]^t \dot{\mathbf{a}} + r\dot{r} = 0 .$$

The normal of the canal surface at a point $\mathbf{x}$ of contact with $S(t)$ coincides with the normal $\mathbf{n} = \mathbf{x} - \mathbf{a}$ of $S(t)$.

A comparison with the polar plane

$$[\mathbf{x} - \mathbf{a}]^t [\mathbf{y} - \mathbf{a}] = r^2$$

of $\mathbf{y}$ with respect to the sphere $S(t)$ shows that in the limit the radical plane coincides with the polar plane of the pole

$$\mathbf{y} = \mathbf{a} - \dot{\mathbf{a}}\frac{r}{\dot{r}} .$$

This has the following geometric meaning. The canal surface contacts the sphere $S(t)$ in a circle, and the common tangent planes in these points meet in $\mathbf{y}(t)$ which lies on the tangent of $\mathbf{a}(t)$ at t. Moreover, the silhouette outline of the canal surface in the plan is swept out by the outlines of the spheres while the common tangents meet in the projection of $\mathbf{y}$, as illustrated in Figure 24.14. In the case where $\dot{r}$ vanishes, $\mathbf{y}$ lies at infinity and the intersection plane is a diametric plane of S.

Any (planar) intersection of a canal surface is enveloped by the (planar) intersections of the spheres generating the canal surface.

A parametric representation of the canal surface can be derived as follows: Let $\mathbf{a}; \mathbf{a}_1, \mathbf{a}_2, \mathbf{a}_3$ be a local Cartesian system depending continuously on t with $\dot{\mathbf{a}} = \mathbf{a}_1 \alpha$. Then the circle $\mathcal{C}$ of contact has the center

$$\mathbf{c} = \mathbf{a} - \frac{r\dot{r}}{\alpha}\mathbf{a}_1$$

and the radius $\varrho = r\sqrt{a^2 - \bar{r}^2}/\alpha$. Hence, $\mathcal{C}$ has the parametric representation

$$\mathbf{x} = \mathbf{c} + [\mathbf{a}_2\cos\varphi + \mathbf{a}_3\sin\varphi]\varrho \ .$$

With t varying this describes the canal surface.

Note that the great circle of S,

$$\mathbf{z} = \mathbf{a} + [\mathbf{a}_2\cos\varphi + \mathbf{a}_3\sin\varphi]r \ ,$$

does not sweep out a canal surface, in general.

Instead of $\mathbf{a}_1, \mathbf{a}_2, \mathbf{a}_3$, one can use the Frenet frame of $\mathbf{a}(t)$. If $\dot{\mathbf{a}}$ is not parallel to the z-direction, i.e., if $\dot{\mathbf{a}} = [a \quad b \quad c]^t$ and $d^2 = a^2 + b^2 \neq 0$, one can also use the simpler frame

$$\mathbf{a}_1 = \dot{\mathbf{a}}/\alpha, \qquad \mathbf{a}_2 = [-b \quad a \quad 0]^t/d, \qquad \mathbf{a}_3 = [ac \quad bc \quad -d^2]^t/\alpha d \ .$$

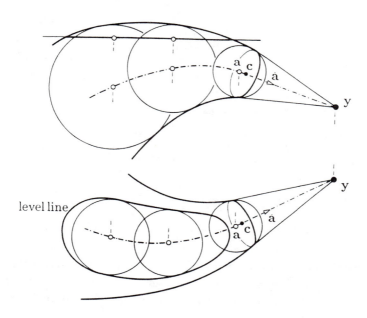

Figure 24.14: Outlines and planar intersection of a canal surface.

Remark 3: Notice that every **surface of revolution** is a canal surface where the center **a** runs on a straight line. Moreover, Dupin's cyclides are also canal surfaces.

24.7 The Four-Dimensional Space

The four-dimensional space can be represented in the same way as the three-dimensional space. Let a point **x** be given by its Cartesian coordinates x, y, z, t. Its orthogonal projections into the tx-plane and the yz-plane are given by

$$\mathbf{x}' = [t \quad x]^t \ , \ \text{and} \ \mathbf{x}'' = [y \quad z]^t \ ,$$

respectively. Any pair, **x**′ and **x**″, corresponds to exactly one point **x** in the 4-space. It is convenient to position both projections as in Figure 24.15. Note that the xy-plane and the zt-plane lend themselves as additional projection planes.

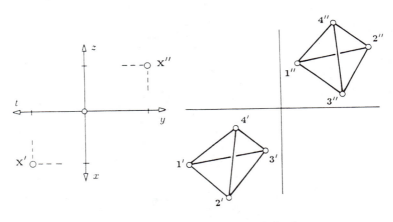

Figure 24.15: Associated projections of the 4-space.

In general, such projection pairs of two points define a line, of three points a plane, and of four points a hyperplane, as illustrated in Figure 24.15. However, a pair of parallelograms defines a parallelogram in 4-space which spans only a plane.

One can rotate each projection in its projection plane. This corresponds to a rotation of the object and also changes the adjacent projections. For example, if one rotates in the xt-plane, the tz- and xy-projection change. This 4D procedure is a tool analogous to the construction of a side elevation in 3D.

Example 5: Consider a hyperplane $\mathcal{U}$ and the points $\mathbf{x}$ of $\mathcal{U}$ which are projected onto some fixed $\mathbf{x}'$. These points lie on a line $\mathcal{B}$ of $\mathcal{U}$ projected into a line $\mathcal{B}''$. Similarly there is a line $\mathcal{A}'$ corresponding to some fixed $\mathbf{x}''$. Rotating the xt- and yz-projection of $\mathcal{U}$ such that $\mathcal{A}'$ and $\mathcal{B}$ become parallel to the x- and y-axes, respectively, the tz-projection of $\mathcal{U}$ becomes a line $\mathcal{C}$. Then rotating $\mathcal{C}$ such that it becomes parallel to the z-axis gives the result shown in Figure 24.16 at the right.

Remark 4: In most cases these special positions simplify the solutions of problems.

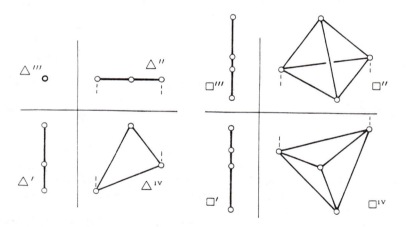

Figure 24.16: Special positions.

Remark 5: Note that in Figure 24.16 the xy- and the yz-projections represent ordinary 3D plans and elevations.

24.8 Notes and Problems

1 A curve in space may be given parametrically or as the intersection of two or more surfaces. The intersection is relatively simple to construct if the surfaces are cylinders. Figure 24.17 shows the curve $\mathbf{x} = [t \quad t^2 \quad t^3]^t$ which lies on the three cylinders

$$y^3 = z^2, \qquad x^3 = z, \qquad x^2 = y \ .$$

The curve is also the intersection of the cone $xz = y^2$ and the hyperbolic paraboloid $xy = z$ which also share the line $y = z = 0$.

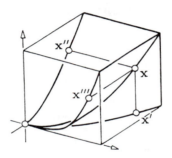

Figure 24.17: The intersection of cylinders.

2 Any parallel projection of a planar figure into a parallel plane can be viewed as a translation which does not change the shape of the figure.

3 The plan of a right angle is a right angle if one of its legs is horizontal.

4 A plan projection of a sloping line segment is shorter than the segment itself. The projection has the same length if and only if it is parallel to the plan.

5 Consider the family of diameters of a circle and their plan projection. As a consequence of Note 4 there is exactly one diameter whose projection equals the diameter of the circle. It coincides with the principal axis of the ellipse.

6 Any plane figure can be rotated around a horizontal line of its plane into a position parallel to the plan. The relationship between the plan of both figures is called a **compression**. It is an axial affinity where the join of any two related points is orthogonal to the axis.

7 In order to get the true shape of a planar figure one can realize it for three points and use the affinity defined by the images and preimages of these three points.

8 Note that the contour of a surface can have discontinuities or corners although the silhouette outline is smooth. Figure 24.18 shows two examples.

9 The offset of a canal surface is a canal surface.

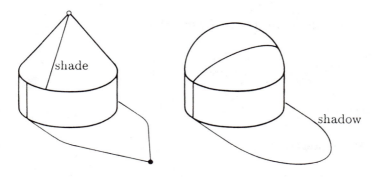

Figure 24.18: Contour and outline, examples.

25 Penetrations

Besides the construction of curves and surfaces, descriptive geometry encompasses the investigation of surface penetrations and, in particular, methods of constructing points and tangents of intersection curves. In order to develop such methods, while minimizing the constructional and the computational effort, it is helpful to know the possible shapes of penetrations and whether they are composed of simpler types.

Literature: Lighton, Rehbock, Wunderlich

25.1 Intersections

The principle of constructing the intersection of a curve and a surface or the intersection of two surfaces is quite simple if one uses an auxiliary surface, as illustrated in Figure 25.1:

Curve and surface: Take an auxiliary surface which contains the given curve and whose intersection with the given surface is simple to handle. Then find the intersection of both curves. This is the desired point.

Surface and surface: Take an auxiliary surface whose intersection curves with both given surfaces are simple to handle. Then find the intersection of these curves. Moving the auxiliary surface generates a series of points on the desired intersection curve.

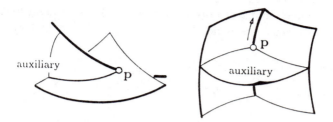

Figure 25.1: The construction principle.

The practicality of these techniques is mainly determined by the choice of the auxiliary surface. Often planes or spheres are chosen for simplicity. Notice, however, that the auxiliary surface itself may be complicated; only its intersection curves with the given surfaces should be simple, i.e., simple to deal with. Obviously, straight lines or circles are most desirable as these intersection curves.

The surface-surface intersection algorithm above generates a sequence of points. However, a point with a tangent is better than two points. A tangent of the intersection curve is, in general, the intersection of the respective tangent planes, as illustrated in Figure 25.2, while a tangent plane of a surface is spanned by two tangents. These facts can be used to find the tangents of an intersection curve.

Figure 25.2: Tangent plane and tangent.

Example 1: In Figure 25.3, the penetration of two cylinders is sketched using only a few points and tangents. Suitable auxiliary surfaces are horizontal planes which intersect both cylinders in pairs of straight lines. Note that only one branch of the penetration of both complete cylinders is shown.

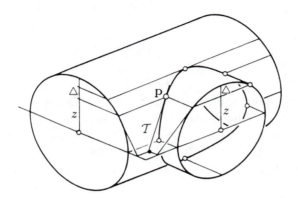

Figure 25.3: Intersection of two cylinders.

25.2 Distinguished Points

Often, points of a penetration curve which are in some distinguished position can be found or constructed in a straightforward way. In many cases the tangents at distinguished points are also quite obvious. A distinguished position can mean a point in extremal position with respect to the object or with respect to the object's position in space. It can also mean points in symmetry planes or on the contour. For example, extremal points are used in Figure 25.3.

Example 2: Figure 25.4 shows some distinguished points of the intersection of two cylinders having a point of contact **d**.

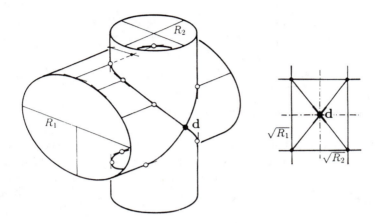

Figure 25.4: Distinguished points with tangents, double point.

25.3 Double Points

If the penetrating surfaces have a **point of contact**, both tangent planes coincide, and in general the point is a **double point** of the intersection curve. Figure 25.4 shows an example.

The construction of the tangents in Section 25.1 fails at such a point. However, there is a simple way to construct both tangents in a double point by intersecting **Dupin's indicatrix**.

The indicatrix of a surface at an ordinary point $\mathbf{p}$ is defined as follows: Translate the tangent plane at $\mathbf{p}$ by an infinitesimally small amount δ and intersect it with the surface. The intersection curve projected orthogonally into the tangent plane and scaled by $1/\sqrt{2\delta}$ is called the **indicatrix** at $\mathbf{p}$. It is a conic section as shown in Section 32.1. From the construction it follows that the tangents of an intersection curve at a double point join opposite intersections of both indicatrices.

Figure 25.5: Dupin's indicatrix, definition and auxiliary figure.

The following planar construction will help get the indicatrix in simple cases. Consider a circle of radius R and a chord at distance δ from some tangent, as illustrated in the right part of Figure 25.5. Let h be half the chord length. Then $h/\sqrt{2\delta} = \sqrt{R - \delta/2}$ which converges to $r = \sqrt{R}$ as δ goes to zero, where $2r$ is the diameter of the indicatrix and R is the radius of curvature of the surface with respect to the same direction.

Often, it is easy to find the extrema of R and thus the indicatrix. Some examples are shown in Figure 25.6: Dupin's indicatrix of a cylinder is a pair of parallel lines, the indicatrix of a sphere is a circle, and the indicatrix of a torus at the point on the equator is an ellipse. The general computation is discussed in Section 32.1. Note that the size of Dupin's indicatrix depends on the scale used.

Example 3: Figure 25.4 shows the intersection of a cylinder which have a point of contact **d**. The indicatrices are two pairs of lines with distances $\sqrt{R_1}$ and $\sqrt{R_2}$, respectively. Their intersections determine points of the tangents at **d**.

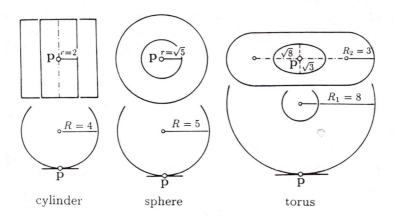

Figure 25.6: Dupin's indicatrix, examples.

25.4 The Order

The **order** of an **algebraic curve** or an **algebraic surface** is defined as the total number of intersection points with an arbitrary plane or straight line, respectively, as illustrated in Figure 25.7. Note that the points can be real, imaginary, multiple, or ideal and that the order is finite.

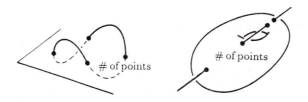

Figure 25.7: The order of a curve and of a surface.

Figure 25.8 shows some examples: Each quadric has order 2, each torus has order 4, and a cubic curve has order 3. A helix and a periodically corrugated infinite sheet of iron are not algebraic curves or surfaces and do not have an order. A curve or surface may be reducible, i.e., it may be composed of several different curves or surfaces of lower degree. Then the order of the curve or surface is the sum of the orders of all individual components. Examples of these decompositions are discussed in Section 25.6.

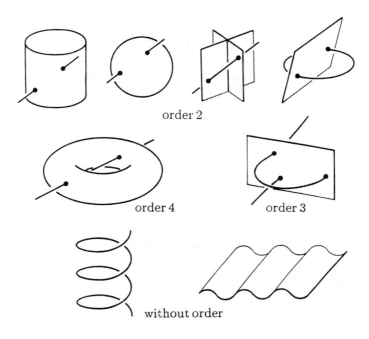

Figure 25.8: The order of curves and surfaces, examples.

Remark 1: If more than n points of an algebraic curve of order n lie in a plane, a component of the curve lies in this plane. Analogously, if more than n points of an algebraic surface of order n lie on a straight line, the line lies completely on the surface.

25.5 Bezout's Theorem

A curve can be given parametrically or as the intersection of surfaces. The following theorems from algebraic geometry provide some useful information on the order of penetrations:

Two surfaces of order m and n without common components intersect in a curve of order $m \cdot n$.

A curve of order n meets a surface of order m not containing any part of the curve in $m \cdot n$ points.

Two curves of a plane of order m and n without common components intersect in $m \cdot n$ points.

The last theorem is known as **Bezout's theorem**. Note that imaginary, multiple, and ideal elements are included. The three theorems are illustrated in Figure 25.9. Note that they reduce to the definition of the order if m or $n = 1$.

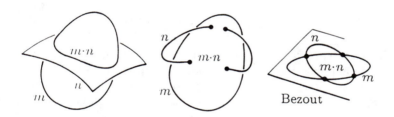

Figure 25.9: The order of intersections.

Example 4: Two quadrics intersect in a curve of order 4. Figures 25.3 and 25.4 show two examples.

Example 5: Any space curve of order 2 is a planar curve. As a consequence, any curve of order 2 on a sphere is a circle.

25.6 Decompositions

A curve can happen to be composed of two or more curves. Then the order
of the curve equals the sum of the orders of its components. For example,
two lines in a plane make up a conic section, a line and an intersecting
conic section form a cubic in the plane or space, two twice intersecting
conic sections make up a curve of order four, etc. Notice that the individual curves intersect each other and that their composition is a curve with
double or multiple points, as illustrated in Figure 25.10. Note further that
these intersection points can be non-real.

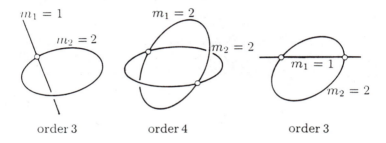

order 3   order 4   order 3

Figure 25.10: Decomposition of curves.

Of particular interest are decompositions of penetration curves since decomposition simplifies their construction. To exploit a decomposition in
practice, simple criteria are needed to decide if a penetration curve is reducible. For example, the penetration of two quadrics is a curve of order
4. It is composed of two conic sections if one of the following equivalent
conditions is met:

1) Both quadrics contain a common conic section.

2) Both quadrics have two points of contact.

3) Both quadrics contact a third quadric, each along a conic section.

The first condition is evident, the other two are closely related to Remark 1
and the discussions in Section 23.6 on pencils and ranges of quadrics. Figure 25.11 shows the following three applications:

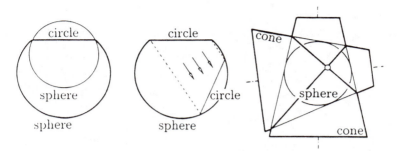

Figure 25.11: A common pair of conic sections.

All spheres contain the ideal spherical circle. Therefore the intersection curve of two spheres is composed of an ideal and a finite (real or imaginary) circle, cf. Example 5.

Parallel light shining through the circular opening of a spherical bowl penetrates the water as a quadratic cylinder. This cylinder has the circle of the opening in common with the the bowl. Therefore the cylinder meets the bowl in a second circle.

Two cones circumscribed to a sphere, as shown at the right, intersect in two ellipses.

Example 6: Any two quadratic cones with a common generator also intersect in a cubic curve, as illustrated in Figure 25.12.

Example 7: The hyperbolic paraboloid $xy = z$ intersects the parabolic cylinder $y = x^2$ in the cubic curve $\mathbf{x} = [t \quad t^2 \quad t^3]^{\mathrm{t}}$ and in the ideal line of the plane $x = 0$. The cubic also lies on the cone $xz = y^2$, which meets the two other surfaces in the x- and the z-axis, respectively.

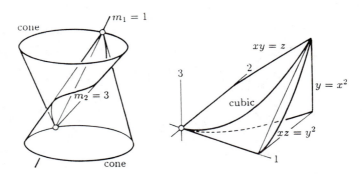

Figure 25.12: Cubic and line.

25.7 Projections

From the definition of the order one can deduce the following. The cone from an arbitrary point c to an algebraic curve of order n has order n. Hence, any projection of the curve from a point not on the curve into a plane is of order n. If c is an ordinary point on the curve, the projection is of order $n-1$.

Any straight line through c, intersecting a spatial curve of order n more than once, corresponds to a multiple point of the projection.

Note that the cone may be a multiple surface and the projected curve a multiple curve. In particular, if all projection rays are chords which meet the curve exactly twice, the curve is of some even order $n = 2m$ while its projection is of order m, but counts twice. Note that complex conjugate points may also have a real projection. Figure 25.13 illustrates these properties.

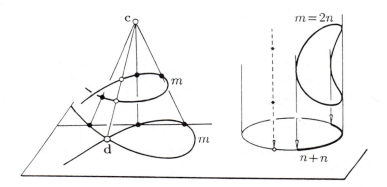

Figure 25.13: Projections of a curve.

Example 8: The penetrations shown in Figure 25.3 and 25.4 are of order four. Their projections in the direction of one of the cylinders into the cylinder base are curves of order two and multiplicity two.

Example 9: A cubic curve, such as in Figure 25.12, is projected from each of its points into a curve of order two.

25.8 Notes and Problems

1 A curve of order 1 is a straight line; a surface of order 1 is a plane.

2 Each plane meets a surface of order n in a curve of order n or lies completely on it.

3 A line meets a space cubic in at most two points.

4 A point where a planar curve changes from one side of its tangent to the other is called an **inflection point**.

5 Consider a spatial curve. An inflection point of its projection corresponds to a point where the osculating plane meets the center of projection.

6 A **torus** has order 4. One gets its equation by rotating the circle $(x-a)^2 + z^2 = r^2$ in the plane $y = 0$ around the z-axis,

$$(x^2 + y^2 + z^2 - a^2 - r^2)^2 = 4a^2(r^2 - z^2) \ .$$

7 Every plane through the axis of a torus intersects the torus in two circles which intersect each other in the circular points of the plane. The following three notes are a consequence of this fact.

8 Every torus intersects itself in the spherical circle.

9 The finite component of the intersection of a torus and a sphere is a curve of order four.

10 If a sphere has two points of contact with a torus, this curve of order four decomposes into two circles, which can also coincide.

11 Consider two cylinders given by the equations

$$Q_1 = x^2 - 2R_1 + z^2 = 0 \quad \text{and} \quad Q_2 = x^2 - 2R_2 + y^2 = 0 \ .$$

Two such cylinders are shown in Figure 25.4. The penetration curve lies on the cone

$$R_2 Q_1 - R_1 Q_2 = (R_2 - R_1)x^2 - R_1 y^2 + R_2 z^2 = 0 \ .$$

This cone meets the plane $x = 0$ in the lines $\sqrt{R_2}z = \pm\sqrt{R_1}y$ representing the tangents at the double point.

12 Let $z = ax^2 + 2bxy + cy^2$ be the quadratic Taylor polynomial of a surface at the origin. Then the Dupin indicatrix of the surface at the origin is given by

$$ax^2 + 2bxy + cy^2 = \frac{1}{2} \ , \quad z = 0 \ .$$

PART SEVEN

Basic Algebraic Geometry

Algebraic geometry deals with algebraic manifolds in projective spaces. In particular, algebraic geometry is concerned with invariants of curves and surfaces under projective and birational transformations. Isaac Newton (1643–1729) and Leonhard Euler (1707–1783) were the first to investigate algebraic properties of curves, but the actual founder of algebraic geometry is Max Noether (1844–1921). In the 20th century, algebraic geometry grew quickly and became more abstract. The more geometric part of algebraic geometry generalizes the methods developed above for quadrics to curves and surfaces of higher degree.

In 1984 Tom Sederberg introduced methods of algebraic geometry into geometric design when he considered calculating the intersection of algebraic curves and surfaces. Nearly at the same time an interesting relation between polynomial splines and polar forms was discovered by de Casteljau. This discovery lead to new insight into the properties of splines and their algorithmic construction.

Furthermore, the implicitization of parametric curves and surfaces and the so-called inverse problem can be solved by methods of classical algebraic geometry.

26 Implicit Curves and Surfaces

A curve in the plane or a surface in space may be given either paramet-rically, i.e., explicitly, or by an equation for the coordinates of its points, i.e., implicitly. If the curve or surface is given by one or more polynomials it is referred to as an **algebraic curve** or **surface**.

Literature: Brieskorn·Knörrer, Griffiths·Harris, Samuel, Walker

26.1 Plane Algebraic Curves

A polynomial $f(\varkappa)$ of degree $n > 0$ in $\varkappa$ is called **homogeneous of degree** n if for all $\varrho \neq 0$

$$f(\varkappa\varrho) = \varrho^n f(\varkappa) \ .$$

Let $f(\varkappa)$ be such a polynomial in three variables. Then the points $\varkappa\varrho = [x \ \ y \ \ z]^t$ satisfying $f(\varkappa) = 0$ form an implicit **algebraic curve** of degree n which also is denoted by f. Let the affine part of f be given by the inhomogeneous polynomial $f(\mathbf{x}) = f(x, y)$ obtained from $f(\varkappa)$ by setting $z = 1$,

$$\begin{aligned} f(\mathbf{x}) = \ &c_{00} \\ &+ c_{10}x + c_{01}y \\ &+ c_{20}x^2 + c_{11}xy + c_{02}y^2 \\ &+ \cdots \\ &+ c_{n0}x^n + \cdots \qquad\qquad \cdots + c_{0n}y^n \end{aligned}$$

which is abbreviated by

$$f = f_0 + f_1 + \cdots + f_n \; ,$$

where f_k summarizes the terms of proper degree k. If the ideal line $z = 0$ is not a component of f, one has $f_n \neq 0$ and $f(\mathbf{x})$ is of the same degree n as $f(\varkappa)$. In matrices $f(\mathbf{x})$ is written as

$$f(\mathbf{x}) = \begin{bmatrix} 1 & x & \cdots & x^n \end{bmatrix} \begin{bmatrix} c_{00} & \cdots & c_{0n} \\ \vdots & \ddots & \\ c_{n0} & & \end{bmatrix} \begin{bmatrix} 1 \\ y \\ \vdots \\ y^n \end{bmatrix}$$

which is abbreviated by $f = \varkappa_n^t C \mathbf{y}_n$.

A change of the affine system does not change the degree and the triangular appearance of the matrix C. C.

Remark 1: The polynomial f depends on the $\frac{1}{2}(n+1)(n+2)$ homogeneous parameters c_{ij}. As a consequence, the curve $f(\mathbf{x}) = 0$ is defined by $\frac{1}{2}(n+1)(n+2) - 1 = \frac{1}{2}n(n+3)$ points in generic position. For example, a conic section is generally defined by 5 points and a cubic curve by 9 points.

26.2 Multiple Points

The behavior of the curve at a fixed point $\mathbf{p}$ can be studied by the intersections with straight lines $\mathbf{x}(\lambda) = \mathbf{p} + \mathbf{v}\lambda$ of the pencil with base $\mathbf{p}$. If $\lambda = 0$ is an r-fold root of $f(\mathbf{x}(\lambda)) = 0$ for all lines of the pencil but not an $(r+1)$-fold root for some of the lines, $\mathbf{p}$ is said to have **multiplicity** r. The lines of the pencil where $\lambda = 0$ is at least an $(r+1)$-fold root are called **tangents** at $\mathbf{p}$.

One can check that the definition of the multiplicity does not depend on the coordinate system. Therefore it may be assumed that $\mathbf{p}$ is the origin $\mathbf{o}$. Then one has the following properties of the curve f at $\mathbf{p}$:

If $f_0 = c_{00} = 0$, the origin lies on the curve.

If $f_0 = 0$ but $f_1 \not\equiv 0$, then $f_1 = c_{10}x + c_{01}y = 0$ defines the tangent at $\mathbf{o}$, and $\mathbf{o}$ is a **simple point**, i.e., it has multiplicity one.

If $f_0 = f_1 \equiv 0$ but $f_2 \not\equiv 0$, then $\mathbf{o}$ is a **double point**, and $f_2 = c_{20}x^2 + c_{11}xy + c_{02}y^2 = 0$ defines a pair of lines which are the tangents of the two branches of the curve at $\mathbf{o}$.

Similarly, if $f_0 = f_1 = f_2 \equiv 0$ but $f_3 \not\equiv 0$, the origin is a **triple point**, where $f_3 = 0$ defines three tangents corresponding to the roots of $f_3 = 0$.

Finally, if $f_0 = \cdots = f_{n-1} \equiv 0$ but $f_n \not\equiv 0$, the curve degenerates to the union of n lines meeting at $\mathbf{o}$.

Notice that the tangents of the curve at any point $\mathbf{p}$ can be real, coalescing, or pairwise conjugate complex although $\mathbf{p}$ is real. Figure 26.1 shows some examples.

Note also that $f_n = 0$ determines the asymptotic directions to the ideal points of $f = 0$.

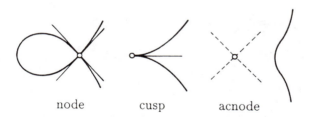

node   cusp   acnode

Figure 26.1: Tangents at a double point.

Remark 2: The **genus** of a plane algebraic curve is defined as

$$\text{genus} = \frac{1}{2}(n-1)(n-2) - (\text{number of double points}) .$$

The double points have to be counted after all finite and ideal multiple points are   resolved by a "blowing up" procedure, as illustrated by the Examples in Figure 26.2.

cusp   triplepoint

Figure 26.2: Resolving multiple points.

26.3 Euler's Identity

From the inhomogeneous polynomial $f(\mathbf{x})$ one can get the homogeneous polynomial $f(\mathbf{x})$ back simply by

$$f(\mathbf{x}) = f_0 z^n + f_1 z^{n-1} + \cdots + f_n 1 .$$

From

$$f(\mathbf{x}) = \sum_{i,j} c_{ij} x^i y^j z^k , \qquad i + j + k = n ,$$

follows **Euler's identity**

$$f_x x + f_y y + f_z z = nf ,$$

where f_x is the partial derivative of f with respect to x at $\mathbf{x}$, etc.

With the aid of this identity a symmetric form of the tangent at some simple point $\mathbf{x}$ of the curve can be derived. Let $\mathfrak{a} = \begin{bmatrix} a & b & c \end{bmatrix}^t$ be an arbitrary point on the tangent, then the tangent at $\mathbf{x}$ is given by

$$f_x(a - x) + f_y(b - y) + f_z(c - z) = 0 ,$$

and because of Euler's identity it is also given by

$$f_x a + f_y b + f_z c = 0 .$$

Remark 3: Applying Euler's identity to itself one gets

$$(f_{xx}x + f_{xy}y + f_{xz}z)x$$
$$+ (f_{xy}x + f_{yy}y + f_{yz}z)y$$
$$+ (f_{xz}x + f_{yz}y + f_{zz}z)z = n(n-1)f .$$

26.4 Polar Forms of Curves

The expression

$$\triangle_\mathbf{o} f = \frac{1}{n}(f_x a + f_y b + f_z c)$$

is called the **first polar form** of f with respect to the **pole** $\mathbf{o} = [a \quad b \quad c]^t$. Obviously, the first polar form is linear in $\mathbf{o}$, and it is a homogeneous polynomial of degree $n-1$ in $\mathbf{x}$. Again, f_x stands for $f_x(\mathbf{x})$, etc.

The polar form has the following properties:

If $\mathbf{o} = \mathbf{x}$, the polar form agrees with $f(\mathbf{x})$.

If $\mathbf{x}$ is fixed on the curve while $\mathbf{o}$ varies, $\triangle_\mathbf{o} f = 0$ defines the tangent of the curve at $\mathbf{x}$.

If $\mathbf{o}$ is fixed while $\mathbf{x}$ varies, $\triangle_\mathbf{o} f = 0$ defines a curve of degree $n-1$ in $\mathbf{x}$, called the **first polar** of $f = 0$ with respect to $\mathbf{o}$.

Together these properties establish the following intriguing geometric meaning of the first polar: All tangents of the curve $f = 0$ at the intersection points with $\triangle_\mathbf{o} f = 0$ meet $\mathbf{o}$ and these are the only tangents from $\mathbf{o}$ to f. If f is free of singularities, it follows from Bezout's theorem that there are $n(n-1)$ such tangents from $\mathbf{o}$ to $f = 0$. Figure 26.3 shows an example where $n = 3$.

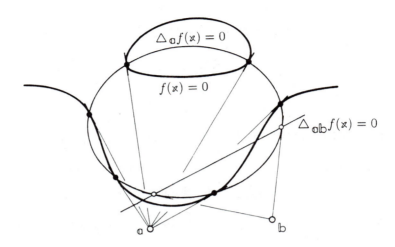

Figure 26.3: First and second polar of a cubic.

Since the first polar form $\triangle_\mathfrak{a} f$ is a polynomial of degree $n - 1$ in $\varkappa$, it has a polar form with respect to any pole $\mathfrak{b}$ itself

$$\triangle_\mathfrak{b}(\triangle_\mathfrak{a} f) = \triangle_{\mathfrak{a}\mathfrak{b}} f \ .$$

This is also illustrated in Figure 26.3. Obviously, this **second polar form** is of degree $n - 2$ in $\varkappa$. One can repeat this construction up to the nth polar form of f with respect to n poles. The nth polar form does not depend on $\varkappa$; it is a point. One can check the following fundamental properties of polar forms:

Polar forms are **linear** in their poles. For $\mathfrak{a} = \mathfrak{p}\alpha + \mathfrak{q}\beta$ one has

$$\triangle_\mathfrak{a} f = \alpha \triangle_\mathfrak{p} f + \beta \triangle_\mathfrak{q} f \ .$$

Polar forms are **symmetric** with respect to their poles,

$$\triangle_{\mathfrak{a}\mathfrak{b}} f = \triangle_{\mathfrak{b}\mathfrak{a}} f \ .$$

Polar forms have **diagonals** which agree with the curve

$$\triangle_x^r f = \underbrace{\triangle_{x \ldots x} f}_{r} = f \ .$$

For $r = 1$, the last property reflects Euler's identity.

26.5 Algebraic Surfaces

Let $F(x)$ be a homogeneous polynomial of degree $n > 0$ in four variables. Then the points $x\varrho = [x \quad y \quad z \quad w]$ satisfying $F(x) = 0$ form an **implicit algebraic surface** in the space which is also denoted by F. Let the affine part of F be given by the inhomogeneous polynomial $F(\mathbf{x}) = F(x, y, z)$ obtained from $F(x)$ by setting $w = 1$,

$$\begin{aligned}
F(\mathbf{x}) = {} & c_{000} \\
& + c_{100}x + c_{010}y + c_{001}z \\
& + \cdots \\
& + c_{n00}x^n + \cdots \qquad\qquad \cdots + c_{00n}z^n
\end{aligned}$$

which is abbreviated by

$$F = F_0 + F_1 + \cdots + F_n \ ,$$

where F_k summarizes the terms of proper degree k. If the ideal plane $w = 0$ is not a component of F, one has $F_n \not\equiv 0$, and $F(\mathbf{x})$ is of the same degree as $F(x)$. A change of the affine system does not change the degree of the surface.

As in the case of curves, one can study the behavior of the surface at a point $\mathbf{p}$ by using the intersection of the surface with the lines $\mathbf{x}(\lambda) = \mathbf{p} + \mathbf{v}(\lambda)$ of the bundle with base $\mathbf{p}$. The multiplicity of $\mathbf{p}$ is defined in exactly the same way and does not depend on the coordinate system. Therefore it suffices to consider the origin $\mathbf{o}$:

If $F_0 = c_{000} = 0$, the origin lies on the surface.

If $F_0 = 0$ and $F_1 \not\equiv 0$, then $F_1 = c_{100}x + c_{010}y + c_{001}z = 0$ defines the tangent plane at $\mathbf{o}$ which is a **simple point** of the surface.

If $F_0 = 0$, $F_1 \equiv 0$ and $F_2 \not\equiv 0$, then the origin is a **double point** and $F_2 = c_{200}x^2 + \cdots + c_{002}z^2 = 0$ represents a cone with apex $\mathbf{o}$. Any generatrix of the cone is a tangent since it meets $F = 0$ at least three times at $\mathbf{o}$. If a generatrix of the cone also lies on $F_3 = 0$, it is called a **principal tangent** since it meets $F = 0$ at least four times at $\mathbf{o}$. The origin itself is said to be **conic, biplanar,** or **uniplanar,** depending on whether the cone $F_2 = 0$ is a proper cone, a pair of planes or a double plane.

Finally, $F_n = 0$ determines the asymptotic directions to the ideal points of $F = 0$.

From the inhomogeneous polynomial $F(\mathbf{x})$ one gets back the homogeneous polynomial $F(\mathbf{x})$ simply by

$$F(\mathbf{x}) = F_0 w^n + F_1 w^{n-1} + \cdots + F_n 1 .$$

Now, Euler's identity reads

$$F_x x + F_y y + F_z z + F_w w = nF .$$

Examples 1-3: Figure 26.4 shows three examples: A conic double point, a self-intersection consisting of biplanar double points, and a surface formed by the tangents of a space curve with uniplanar double points along the generating curve.

Remark 4: The polynomial F depends on the $\frac{1}{6}(n+1)(n+2)(n+3)$ homogeneous parameters c_{ijk}. As a consequence F is defined by $\frac{1}{6}(n+1)(n+2)(n+3) - 1$ points in generic position. For example, a quadric is generally defined by 9 points.

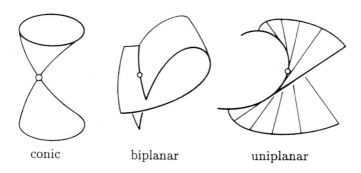

conic   biplanar   uniplanar

Figure 26.4: Double points on surfaces.

26.6 Polar Forms of Surfaces

The expression

$$\triangle_{\mathbb{O}}F = \frac{1}{n}(F_x a + F_y b + F_z c + F_w d)$$

is called the **first polar form** of F with respect to the **pole** $\mathbb{o} = [a \ b \ c \ d]^t$.
Obviously, the first polar form is linear in $\mathbb{o}$ and a homogeneous polynomial
of degree $n-1$ in $\varkappa$. Again, F_x stands for $F_x(\varkappa)$, etc.

The polar form has the following properties:

If $\mathbb{o} = \varkappa$, the polar form agrees with $F(\varkappa)$.

If $\varkappa$ is fixed on the surface while $\mathbb{o}$ varies, $\triangle_{\mathbb{O}}F = 0$ defines the tangent
plane of the surface at $\varkappa$.

If $\mathbb{o}$ is fixed while $\varkappa$ varies, $\triangle_{\mathbb{O}}F = 0$ defines a surface of degree $n-1$,
called the **first polar** of $F = 0$ with respect to $\mathbb{o}$.

Together these properties establish the following intriguing geometric meaning of the first polar: The tangent planes of the surface $F = 0$ at the inter-
section points with the polar $\triangle_{\mathbb{O}}F = 0$ meet $\mathbb{o}$ as illustrated in Figure 26.5,
and these are the only tangent plane from $\mathbb{o}$ to F.

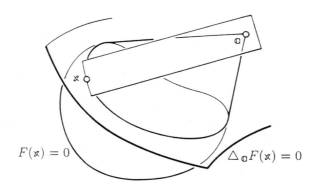

$$F(x) = 0 \qquad\qquad \triangle_\mathrm{o}F(x) = 0$$

Figure 26.5: First polar of a surface.

Since the first polar form $\triangle_\mathrm{o}F$ is a polynomial of degree $n-1$ in x, it may again be subject to the construction of a polar form with respect to a pole b,

$$\triangle_\mathrm{b}(\triangle_\mathrm{o}f) = \triangle_\mathrm{ob}f \ .$$

Obviously, this **second polar form** is of degree $n-2$ in x. One can repeat this construction up to the nth polar form of F with respect to n poles. The nth polar form does not depend on x; it is a point. Analogous to the polar forms of curves, one has the following fundamental properties:

Polar forms are **linear** in their poles, i.e., for $\mathrm{o} = \mathrm{p}\alpha + \mathrm{q}\beta$ one has

$$\triangle_\mathrm{o}F = \alpha\,\triangle_\mathrm{p}F + \beta\,\triangle_\mathrm{q}F \ .$$

Polar forms are **symmetric** with respect to their poles,

$$\triangle_\mathrm{ob}F = \triangle_\mathrm{bo}F \ .$$

Polar forms have **diagonals** which agree with the surface,

$$\triangle_x^r F = \triangle_{\underbrace{x \ldots x}_{r}}F = F \ .$$

For $r = 1$, the last property reflects Euler's identity.

26.7 Notes and Problems

1 Euler's identity can also be generalized to homogeneous polynomials of more variables.

2 As a consequence of Bezout's theorem, any plane cubic curve with two double points is composed of a conic and the line joining the double points.

3 Let f be a plane curve of degree n without multiple components and let multiple, ideal and complex points be properly counted. Then one has:

Any line has exactly n intersections with f. Hence, the order of f is n.

Any line through a point $\mathbb{p}$ of multiplicity r intersects f in $n - r$ further points. Moreover, there is a line through $\mathbb{p}$ where these $n - r$ points are distinct.

4 The number of tangents of a non-degenerate plane algebraic curve which lie in a pencil is called the **class** of the curve.

5 The class of a curve of order n without multiple points is $m = n(n - 1)$. One has $m = n$ only for (non-degenerate) conic sections.

6 A finite sequence of projective transformations and quadratic transformations $\mathbb{x} \to \mathbb{y} = [1/x \quad 1/y \quad 1/z]^t$ can be used to transform any irreducible curve into a curve whose multiple points are only ordinary double points.

7 The transformations of Note 6 do not change the genus of a curve.

8 A planar algebraic curve f can be represented as a rational parametric curve only if its genus is 0.

9 The degree and order of an algebraic surface F are equal.

10 The number of tangent planes of a non-degenerate algebraic surface which lie in a pencil is called the **class** of the surface.

11 The class of an algebraic surface of order n without multiple points is $m = n(n - 1)^2$. One has $m = n$ only for (non-degenerate) quadrics.

12 Obviously, class and order are dual concepts.

27 Parametric Curves and Surfaces

Parametric curves and surfaces are quite popular since they permit direct numerical control for design and manufacturing and they are easy to work with. Of particular interest are polynomial curves and surfaces. Such curves and surfaces are algebraic. Moreover, splines are composed of polynomial segments and are strongly related to the so-called osculants which are associated curves of lower degree.

Literature: Boehm·Farin·Kahmann, Farin, Hoffmann, Liu

27.1 Rational Curves

A parametric curve in the affine space $\mathcal{A}^d$ is defined by

$$\mathbf{x}(t) = \begin{bmatrix} x_1(t) \\ \vdots \\ x_d(t) \end{bmatrix} ,$$

where the $x_i(t)$ are arbitrary functions of the parameter t. In particular, if these functions are polynomials of degree n, the curve is called a **polynomial curve**. Then it has a representation of the form

$$\mathbf{x}(t) = \mathbf{a}_0 + \mathbf{a}_1 t + \cdots + \mathbf{a}_n t^n , \qquad \mathbf{x}, \mathbf{a}_i \in \mathbb{R}^d$$

or in matrix notation

$$\mathbf{x}(t) = \mathbf{a}_0 + [\,\mathbf{a}_1 \ldots \mathbf{a}_n\,] \begin{bmatrix} t \\ \vdots \\ t^n \end{bmatrix}$$

which is abbreviated by $\mathbf{x}(t) = \mathbf{a}_0 + A\mathbf{t}_n$. The curve $\mathbf{t}_n = [t \ldots t^n]^t$ is called the **normal curve** in $\mathcal{A}^n$. It follows that any polynomial curve of degree n in $\mathcal{A}^d$ can be viewed as the affine image of the normal curve in $\mathcal{A}^n$, as illustrated in Figure 27.1. Note that the normal curve of degree n spans $\mathcal{A}^n$.

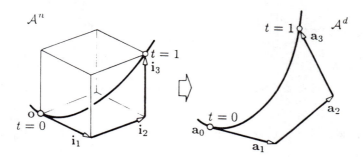

Figure 27.1: The cubic normal curve and its affine image.

A **rational curve** of degree n in $\mathcal{A}^d$ is defined by

$$\mathbf{x}(t) = \frac{\mathbf{a}_0 + \mathbf{a}_1 t + \cdots + \mathbf{a}_n t^n}{a_0 + a_1 t + \cdots + a_n t^n} \,,$$

in homogeneous coordinates by

$$\mathbf{x}(t)\varrho = \begin{bmatrix} x_0(t) \\ \mathbf{x}(t) \end{bmatrix} = \begin{bmatrix} a_0 \\ \mathbf{a}_0 \end{bmatrix} + \begin{bmatrix} a_1 \\ \mathbf{a}_1 \end{bmatrix} t + \cdots + \begin{bmatrix} a_n \\ \mathbf{a}_n \end{bmatrix} t^n \,,$$

and in matrices by

$$\mathbf{x}(t)\varrho = \begin{bmatrix} a_0 & \cdots & a_n \\ \mathbf{a}_0 & \cdots & \mathbf{a}_n \end{bmatrix} \begin{bmatrix} 1 \\ t \\ \vdots \\ t^n \end{bmatrix}$$

which is abbreviated by $\varkappa(t)\varrho = \mathbb{A}\mathfrak{t}_n$ where $\mathfrak{t}_n = [1 \quad t \ldots t^n]^t$ represents
the normal curve in extended coordinates. It follows that any rational curve
of degree $\leq n$ in $\mathcal{P}^d$ can be viewed as a projective image of the normal
curve $\mathfrak{t}_n$ in the projective extension of $\mathcal{A}^n$, as illustrated in Figure 27.2.

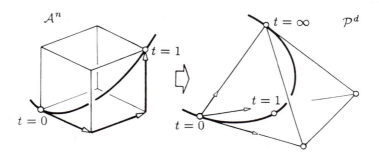

Figure 27.2: The cubic normal curve and its projective image.

Remark 1: On homogeneizing t by $t = t_1/t_0$ one gets $\mathfrak{t}_n\tau = [t_0^n \ldots t_1^n]$.

27.2 Changing the Parameter

Any **affine** map of the parameter,

$$s = a + bt ,$$

does not change a polynomial curve as a whole. However, it changes the
parameter values of the points on the curve. Moreover, the reparametriza-
tion induces an affine map $\mathbf{y} = \mathbf{b} + B\mathbf{x}$ of $\mathcal{A}^n$ onto itself which maps the
normal curve onto itself such that the points $\mathbf{t}_n$ are mapped onto the points
$\mathbf{s}_n$,

$$\mathbf{s}_n = \mathbf{b} + B\mathbf{t}_n .$$

More generally, for an arbitrary parametric curve $\mathbf{x} = \mathbf{a} + A\mathbf{t}_n$ spanning
$\mathcal{A}^n$, this reparametrization induces the affine map

$$\mathbf{y} = \mathbf{a} + A[\mathbf{b} + BA^{-1}[\mathbf{x} - \mathbf{a}]]$$

which maps $\mathbf{x}(t)$ onto $\mathbf{x}(s)$.

Similarly, a **projective** map of the parameter,

$$s = \frac{s_1}{s_0} = \frac{a + bt}{c + dt} , \quad t = \frac{t_1}{t_0} ,$$

or in homogeneous form

$$\begin{bmatrix} s_0 \\ s_1 \end{bmatrix} = \begin{bmatrix} c & d \\ a & b \end{bmatrix} \begin{bmatrix} t_0 \\ t_1 \end{bmatrix} ,$$

does not change a rational curve as a whole. However, it changes the parameter values of the points on the curve.

Moreover, the reparametrization induces a collineation $y\sigma = \mathbb{B}\mathfrak{x}$ of $\mathcal{P}^n$ onto itself which maps the normal curve onto itself such that the points $\mathfrak{t}_n$ are mapped onto the points $\mathfrak{s}_n$,

$$\mathfrak{s}_n = \mathbb{B}\mathfrak{t}_n .$$

More generally, for any parametric curve $\mathfrak{x} = \mathbb{P}\mathfrak{t}_n$ spanning $\mathcal{P}^n$, this reparametrization induces the projective map

$$y\sigma = \mathbb{P}\mathbb{B}\mathbb{P}^{-1}\mathfrak{x}$$

which maps $\mathfrak{x}(t)$ onto $\mathfrak{x}(s)$.

Example 1: O'Neill's parabola $z^2 = y^3$ in the yz-plane can be viewed as the elevation of the normal curve $\mathbf{t}_3 = [\, t \quad t^2 \quad t^3 \,]^{\mathbf{t}}$ in $\mathcal{A}^3$, as illustrated in Figure 27.3. The simple shift of the parameter $s = t + 1$ induces the affine map

$$\mathbf{y} = \begin{bmatrix} 1 \\ 1 \\ 1 \end{bmatrix} + \begin{bmatrix} 1 & & \\ 2 & 1 & \\ 3 & 3 & 1 \end{bmatrix} \mathbf{x}$$

which maps the normal curve onto itself. Of course, the elevation of the curve is also left unchanged as a whole. However, there does not exist any affine map of the yz-plane which maps O'Neill's parabola onto itself.

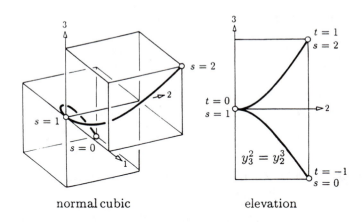

normal cubic   elevation

Figure 27.3:   O'Neill's parabola.

27.3   Osculants of a Curve

Consider a polynomial curve $\mathbf{x}(t)$ of proper degree n in $\mathcal{A}^d$ and denote its derivative with respect to t by $\dot{\mathbf{x}}$. Then for each fixed a the point

$$\mathbf{y} = \mathbf{x} + \dot{\mathbf{x}} \cdot (a - t)/n$$

traces out a curve of degree $n - 1$ in t. This curve is called the **first osculant** of $\mathbf{x}(t)$ at the knot a and denoted by $\Delta_a \mathbf{x}$, see Figure 27.4 for an illustration.

The osculant has the following nice properties:

If t is kept fixed while a varies, $\mathbf{y}$ traces out the tangent of $\mathbf{x}$ at t.

If a is kept fixed while t varies, $\mathbf{y}$ traces out the osculant of $\Delta_a \mathbf{x}$.

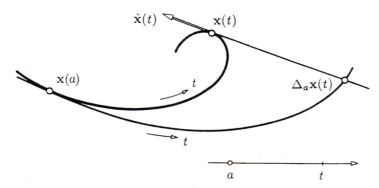

Figure 27.4: The first osculant of a curve.

At $t = a$, one has $\overset{i}{\mathbf{y}} \cdot n = \overset{i}{\mathbf{x}} \cdot (n - i)$, $i = 0, \ldots, n - 1$, where the superscript i denotes the ith derivative with respect to t.

Let the span of $\dot{\mathbf{x}}, \ldots, \overset{s}{\mathbf{x}}$ be of dimension r, then the span of $\mathbf{x}; \dot{\mathbf{x}}, \ldots, \overset{s}{\mathbf{x}}$ is called the rth **osculating flat** of $\mathbf{x}$ at t. Thus, the last property above means that the osculating flats of $\mathbf{x}$ and $\mathbf{y}$ agree at a.

Since the osculant $\Delta_a\mathbf{x}$ is a polynomial curve of degree $n - 1$ in t, it has an osculant at any knot b itself,

$$\Delta_{ab}\mathbf{x} = \Delta_b(\Delta_a\mathbf{x})$$
$$= \mathbf{x} + \dot{\mathbf{x}} \cdot (a - 2t + b)/n + \ddot{\mathbf{x}} \cdot (a - t)(b - t)/n(n - 1) \ .$$

The osculant $\Delta_{ab}\mathbf{x}$ is symmetric with respect to the knots a and b, and it is called the **second osculant** of $\mathbf{x}(t)$ at a and b. This procedure can be repeated up to the nth osculant of $\mathbf{x}(t)$ with respect to n knots; this last osculant is a point.

Osculants have the following further properties:

They are **affine** in their knots: For $a = p\alpha + q\beta$, $\alpha + \beta = 1$ one has

$$\Delta_a\mathbf{x} = \Delta_p\mathbf{x} \cdot \alpha + \Delta_q\mathbf{x} \cdot \beta \ .$$

They are **symmetric** with respect to their knots,

$$\Delta_{ab}\mathbf{x} = \Delta_{ba}\mathbf{x} \ .$$

Their **diagonals** agree with the curve,

$$\Delta_t^r\mathbf{x} = \Delta_{\underbrace{t \ \ldots \ t}_{r}}\mathbf{x} = \mathbf{x}(t) \ .$$

Let $\mathbf{x}(t)$ span $\mathcal{A}^n$, then the osculant at some knot a lies in the $(n-1)$th osculating flat at a. Hence, the nth osculant represents the intersection of the n osculating hyperplanes at the n knots.

Remark 2: The properties of osculants and polar forms are interrelated. Namely, the coordinates of $\Delta_a\mathbf{x}$ are the polar forms of the coordinates of $\mathbf{x}$ with respect to the pole a:

Recall that the polar form of the homogeneized polynomial $\mathbf{x}(t, w)$ with respect to $\mathfrak{o} = [a \ \ 1]^t$ is given by

$$\Delta_{\mathfrak{o}}\mathbf{x} = \frac{1}{n}[\mathbf{x}_t a + \mathbf{x}_w 1] \ .$$

Using Euler's identity $\mathbf{x}_t t + \mathbf{x}_w w = n\mathbf{x}$ at $[t \ \ 1]^t$, one obtains

$$\Delta_{\mathfrak{o}}\mathbf{x} = \mathbf{x} + \mathbf{x}_t \cdot (a - t)/n = \Delta_a\mathbf{x} \ .$$

Remark 3: One can dispense with the stipulation that $\mathbf{x}(t)$ must be of proper degree n. In the case where $\mathbf{x}(t)$ is of degree $r < n$, the first $n - r$ osculants are only of degree r in t but still affine in each knot, cf. Examples 2 and 4.

Remark 4: If $\mathbf{x}(t)$ is a linear combination of some basis functions $N_i(t)$

$$\mathbf{x}(t) = \sum_i \mathbf{d}_i N_i(t) \ ,$$

one has

$$\Delta_a\mathbf{x}(t) = \sum_i \mathbf{d}_i \Delta_a N_i(t) \ .$$

Example 2: Let $\mathbf{x}(t)$ be some cubic,

$$\mathbf{x}(t) = \mathbf{a}_0 + 3\mathbf{a}_1 t + 3\mathbf{a}_2 t^2 + \mathbf{a}_3 t^3 \ ,$$

then

$$\Delta_a \mathbf{x}(t) = \mathbf{a}_0 + 2\mathbf{a}_1 t + \mathbf{a}_2 t^2 + [\mathbf{a}_1 + 2a_2 t + \mathbf{a}_3 t^2]a \ ,$$
$$\Delta_{ab}\mathbf{x}(t) = \mathbf{a}_0 + \mathbf{a}_1 t + [\mathbf{a}_1 + \mathbf{a}_2 t](a + b) + [\mathbf{a}_2 + \mathbf{a}_3 t]ab \ ,$$
$$\Delta_{abc}\mathbf{x} = \mathbf{a}_0 + \mathbf{a}_1(a + b + c) + \mathbf{a}_2(ab + bc + ca) + \mathbf{a}_3 abc \ .$$

Note that $\Delta_{ab}\mathbf{x}$ is of degree 2 in t, even if $\mathbf{a}_3 = \mathbf{o}$, provided that $\mathbf{a}_2 \neq \mathbf{o}$. A recursive construction of $\Delta^3_{abc}\mathbf{x}$ in t is given in Section 29.7(3).

27.4 Bézier Curves

Since osculants are symmetric and affine in their knots,

$$\Delta_t = \alpha \cdot \Delta_a + \beta \cdot \Delta_b \quad \text{for} \quad t = \alpha a + \beta b \, , \alpha + \beta = 1 \ ,$$

a polynomial curve $\mathbf{x}(t)$ of degree n can be recursively computed from some suitable set of $n + 1$ of its nth osculants. In particular, let

$$\Delta^i_a \, \Delta^j_b \, \mathbf{x} \, , \quad i + j = n \, , \quad i = 0, \dots, n$$

be given. Then one can compute the osculants $\Delta^i_a \, \Delta^j_b \, \Delta^r_b \, \mathbf{x}$ at $t = \alpha a + \beta b$, $\alpha + \beta = 1$, by the recursion

$$\Delta^i_a \, \Delta^j_b \, \Delta^r_t \, \mathbf{x} = \alpha \cdot \Delta^{i+1}_a \, \Delta^j_b \, \Delta^{r-1}_t \, \mathbf{x} + \beta \cdot \Delta^i_a \, \Delta^{j+1}_b \, \Delta^{r-1}_t \, \mathbf{x} \ ,$$

$r = 1, \dots, n, i = 0, \dots, n - r$. Finally, this gives

$$\Delta^n_t \, \mathbf{x} = \mathbf{x}(t) \ .$$

A comparison shows that the recursion coincides with the algorithm of de Casteljau from Section 12.3, where $\mathbf{b}^r_i = \mathbf{p}^r_i$ and $\mathbf{b}^r_i = \mathbf{b}^{r-1}_{i+1}\alpha + \mathbf{b}^{r-1}_i\beta$. Consequently, the given points are the Bézier points

$$\Delta^i_a \, \Delta^j_b \, \mathbf{x} = \mathbf{b}^0_i \ ,$$

and the curve $\mathbf{x}(t)$ has the **Bézier representation**

$$\mathbf{x}(t) = \sum_{i=0}^{n} \mathbf{b}_i^0 B_i^n(u) , \quad t = a(1-u) + bu .$$

Moreover, one can choose a different $t = t_r = \alpha_r a + \beta_r b$ at each level r of the recursion. Then one obtains

$$\Delta_a^i \, \Delta_b^j \, \Delta_{t_1 \, \ldots \, t_r} \mathbf{x} = \mathbf{b}_i^r(t_1 \ldots t_r) .$$

This has the following meaning:

The points $\mathbf{b}_0^r, \ldots, \mathbf{b}_r^r$ control the rth osculant

$$\Delta_{t_1 \, \ldots \, t_r} \mathbf{x}(t) = \sum_{i=0}^{r} \mathbf{b}_i^r(t_1 \ldots t_r) B_i^{n-r}(u)$$

of $\mathbf{x}$ with respect to the knots $t_1, \ldots, t_r$.

Figure 27.5 illustrates this property.

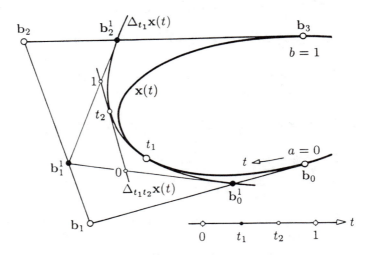

Figure 27.5: Osculants of a cubic Bézier curve.

Example 3: For a cubic curve with Bézier points $\mathbf{b}_0, \ldots, \mathbf{b}_3$ one has

$$\Delta_a \mathbf{x} = \Big[\mathbf{b}_0(1-t)^2 + \mathbf{b}_1 2(1-t)t + \mathbf{b}_2 t^2\Big](1-a)$$
$$+ \Big[\mathbf{b}_1(1-t)^2 + \mathbf{b}_2 2(1-t)t + \mathbf{b}_3 t^2\Big]a \ ,$$

$$\Delta_{abc} \mathbf{x} = \mathbf{b}_0 (1-a)(1-b)(1-c)$$
$$+ \mathbf{b}_1 \Big[(1-a)(1-b)c + (1-a)b(1-c) + a(1-b)(1-c)\Big]$$
$$+ \mathbf{b}_2 \Big[(1-a)bc + a(1-b)c + ab(1-c)\Big]$$
$$+ \mathbf{b}_3 \ abc \ .$$

27.5 Splines

There is an analogous relationship between osculants and the B-spline representation of a polynomial curve $\mathbf{x}(t)$. Namely, let $a_1, \ldots, a_{2n}$ be a sequence of initial knots such that $a_i < a_j$ for $i \le n < j$ and let the $n+1$ points

$$\mathbf{c}_i^0 = \Delta_{a_{i+1} \ldots a_{i+n}} \mathbf{x} \ , \quad i = 0, \ldots, n \ ,$$

be given. Then, for any t one can compute the osculants

$$\mathbf{c}_i^r = \Delta_{a_{i+1} \ldots a_{i+n-r}} \mathbf{x}$$
$$= \Delta_{a_{i+1} \ldots a_{i+n-r}} \Delta_t^r \mathbf{x}$$

by the recursion

$$\mathbf{c}_i^r = \mathbf{c}_{i-1}^{r-1}(1-\alpha_i^r) + \mathbf{c}_i^{r-1}\alpha_i^r \ , \quad \alpha_i^r = \frac{t-a_i}{a_{i+n-r} - a_i} \ ,$$
$$r = 1, \ldots, n; \quad i = r, \ldots, n \ .$$

Finally this gives

$$\mathbf{c}_n^n = \Delta_t^n \mathbf{x} = \mathbf{x}(t) \ .$$

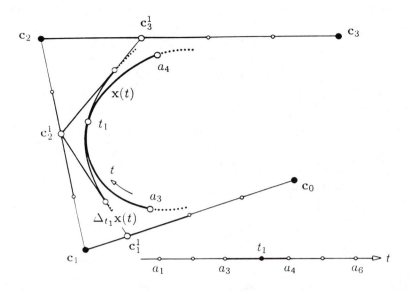

Figure 27.6: First osculant of a cubic spline segment.

This recursive computation is known as **de Boor's algorithm** for computing $\mathbf{x}(t)$, $t \in [a_n, a_{n+1}]$ from the control points $\mathbf{c}_i^0$ of its B-spline representation

$$\mathbf{x}(t) = \sum_{i=0}^{n} \mathbf{c}_i^0 N_i^n(t) ,$$

where $N_i^n(t)$ denotes the B-spline of degree n with respect to the knots $a_i, \ldots, a_{i+n+1}$.

As before, one can choose a different new knot t_r at each level r of de Boor's algorithm and obtains the osculants

$$\underset{a_{i+1}\,\ldots\,a_{i+n-r}}{\Delta}\ \underset{t_1\,\ldots\,t_r}{\Delta}\ \mathbf{x} = \mathbf{c}_i^r .$$

Moreover, one can concludes as before:

The points $\mathbf{c}_{n-r}^r, \ldots, \mathbf{c}_n^r$ are the control points of the B-spline representation of the rth osculant

$$\Delta_{t_1 \ldots t_r} \mathbf{x}(t) = \sum_{i=n-r}^{n} \mathbf{c}_i^r(t_1 \ldots t_r) N_i^{n-r}(t)$$

with respect to the (new) knots $t_1, \ldots, t_r$.

Figure 27.6 shows a cubic spline segment with its first osculant at t_i. Thus, for $r = n$ the points $\mathbf{c}_0^n, \ldots, \mathbf{c}_n^n$ are the control points of the B-spline representation of the polynomial segment $\mathbf{x}(t)$, $t \in [t_n, t_{n+1}]$ with respect to the new knots t_i.

In particular, if the new knots t_i form a refinement of the initial knot sequence a_i, the recursive computation of the $\mathbf{c}_i^n$ is known as the **Oslo algorithm**.

Remark 5: Since $\Delta_a^i \, \Delta_b^{n-i} \, \mathbf{x} = \mathbf{b}_i^0$, the generalized de Boor algorithm can be used to compute the Bézier points of a spline segment.

Remark 6: Since $\Delta_{a_{i+1} \ldots a_{i+n}} \mathbf{x} = \mathbf{c}_i^0$ the generalized de Casteljau algorithm can be used to compute the spline control points of a segment from its Bézier points.

Remark 7: Osculants can also be expressed in terms of symmetric functions as in Example 2. One can also work with this or any other representation of the osculants to compute B-spline control points and the Bézier points.

Example 4: In particular, $x(t) = t$ as a spline function of degree n with respect to the knots $a_1, \ldots, a_{2n}$ has the control points

$$c_i^0 = (a_{i+1} + \cdots a_{i+n})/n$$

which are known as **Greville abscissae**.

27.6 Osculants of a Surface

The idea of osculants can also be transferred to polynomial surfaces $\mathbf{x}(\mathbf{u}) = \mathbf{x}(u, v)$. Let $\mathbf{x}(\mathbf{u})$ be of proper total degree n in $\mathcal{A}^d$. For each fixed point

$\mathbf{a} = [a \quad b]^t,$

$$\Delta_{\mathbf{a}}\mathbf{x} = \Delta_{\mathbf{a}}^1\mathbf{x} = \mathbf{x} + [\mathbf{x}_u(a-u) + \mathbf{x}_v(b-v)]/n$$

is a polynomial of total degree $n-1$ in $\mathbf{u}$. This surface is called the **first osculant** of $\mathbf{x}(\mathbf{u})$ at the **knot a**. Furthermore, analogous to curves, one can form osculants with respect to up to n knots.

These osculants have the same properties as the curve osculants:

They are **affine** in their knots,

$$\Delta_{\mathbf{c}}\mathbf{x} = \Delta_{\mathbf{a}}\mathbf{x} \cdot \alpha + \Delta_{\mathbf{b}}\mathbf{x} \cdot \beta , \quad \mathbf{c} = \mathbf{a}\alpha + \mathbf{b}\beta , \quad \alpha + \beta = 1;$$

they are **symmetric**,

$$\Delta_{\mathbf{ab}}\mathbf{x} = \Delta_{\mathbf{ba}}\mathbf{x};$$

and their **diagonals** represent the surface itself,

$$\Delta_{\mathbf{u}}^r\mathbf{x} = \Delta_{\underbrace{\mathbf{u} \, \ldots \, \mathbf{u}}_{r}}\mathbf{x} = \mathbf{x}(\mathbf{u}) .$$

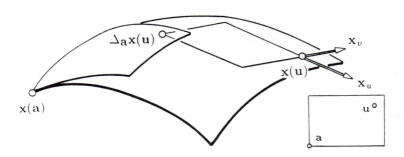

Figure 27.7: The first osculant of a surface.

Furthermore, as is true for curves, there is a relationship between osculants and the Bernstein-Bézier representation of surfaces. Namely let the points

$$\Delta_{\mathbf{a}}^i \, \Delta_{\mathbf{b}}^j \, \Delta_{\mathbf{c}}^k \, \mathbf{x} = \mathbf{b}_{ijk}^0 , \quad i+j+k = n ,$$

be given. Further, let $\mathbf{u} = \mathbf{a}\alpha + \mathbf{b}\beta + \mathbf{c}\gamma$, $\alpha + \beta + \gamma = 1$. Then one can compute the osculants

$$\Delta_{\mathbf{a}}^i \; \Delta_{\mathbf{b}}^j \; \Delta_{\mathbf{c}}^k \; \Delta_{\mathbf{u}}^r \; \mathbf{x} = \mathbf{b}_{ijk}^r \;, \quad i + j + k + r = n$$

by the recursion

$$\mathbf{b}_{ijk}^r = \mathbf{b}_{(i+1)jk}^{r-1} \cdot \alpha + \mathbf{b}_{i(j+1)k}^{r-1} \cdot \beta + \mathbf{b}_{ij(k+1)}^{r-1} \cdot \gamma \;.$$

Finally, this gives

$$\Delta_{\mathbf{u}}^n \; \mathbf{x} = \mathbf{b}_{000}^n = \mathbf{x}(\mathbf{u}) \;.$$

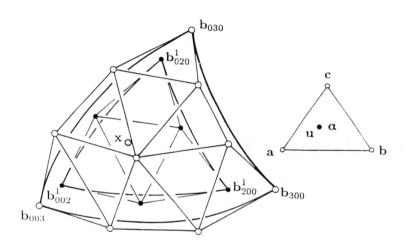

Figure 27.8: The first osculants of a cubic Bézier surface.

This algorithm is known as **de Casteljau's algorithm** for triangular Bézier surfaces. One can check that $\mathbf{x}(\mathbf{u})$ has the Bézier representation

$$\mathbf{x}(\mathbf{u}) = \sum_\Delta \mathbf{b}_{ijk} B_{ijk}^n(\boldsymbol{\alpha})$$

with respect to the triangle $\mathbf{abc}$ as defined in Section 22.8. The $\mathbf{b}_{ijk}$ are the so-called **Bézier points** of the triangular surface $\mathbf{x}$.

As above, one can choose a different knot $\mathbf{u}_r$ at each level r of de Casteljau's algorithm. Then one obtains

$$\Delta_{\mathbf{a}}^i \, \Delta_{\mathbf{b}}^j \, \Delta_{\mathbf{c}}^k \, \Delta_{\mathbf{u}_1 \ldots \mathbf{u}_r} \, \mathbf{x} = \mathbf{b}_{ijk}^r \, , \quad i+j+k+r = n \, .$$

This entails the following:

> The points $\mathbf{b}_{ijk}^r$ control the rth osculant of a Bézier surface $\mathbf{x}(\mathbf{u})$ with respect to the knots $\mathbf{u}_1, \ldots, \mathbf{u}_r$:
>
> $$\Delta_{\mathbf{u}_1 \ldots \mathbf{u}_r} \, \mathbf{x}(\mathbf{u}) = \sum \mathbf{b}_{ijk}^r (\mathbf{u}_1 \ldots \mathbf{u}_r) B_{ijk}^{n-r}(\boldsymbol{\alpha}) \, .$$

Remark 8: Any line $\mathbf{u}(t) = \mathbf{u}_0 + \mathbf{v} \cdot t$ in the parameter domain defines a curve $\mathbf{x}(\mathbf{u}(t))$ on the surface. The first osculant of this curve at the knot $t = t_1$ lies on the first osculant of the surface at the knot $\mathbf{u}_1 = \mathbf{u}_0 + \mathbf{v}t_1$. Moreover,

> if $\mathbf{v}$ varies, but $\mathbf{u}_0$ is kept fix, the curve osculants at $t_0 = 0$ sweep out the surface osculant at $\mathbf{u}_0$.

Remark 9: As is true for curves, the coordinates of the surface osculant $\Delta_{\mathbf{a}} \mathbf{x}$ are the polar forms of the coordinates of $\mathbf{x}$ with respect to the pole in the parameter plane $\mathbf{a}$.

27.7 Notes and Problems

1 A plane polynomial curve of degree n depends on $2(n+1)$ coefficients, while two of them are needed to fix the parametrization. Hence, such a curve depends on $2n$ free coefficients only.

2 A plane rational curve of degree n depends on $3(n+1)$ coefficients, while one is needed for homogeneity and three are needed to fix the parametrization. Hence, such a curve depends on $3n-1$ free coefficients only.

3 Recall that an implicit plane algebraic curve of degree n depends on $\frac{1}{2}(n+1)(n+2)-1$ free coefficients. It is obvious that, even for $n = 3$, the number of implicit curves exceeds the number of rational curves. Hence, there are implicit cubics which do not have a rational representation of degree 3. This statement also holds for curves of degree n with $n > 3$.

4 It can be shown that a planar curve of degree n has a rational parametrization of degree n if and only if it is of genus 0.

5 A non-degenerate intersection of two quadrics in space is not a rational parametric curve.

6 Consider a rational curve or surface represented by homogeneous coordinates $\varkappa(\mathfrak{t})$. If $\varkappa(\mathfrak{a}) = \mathfrak{o}$ for some $\mathfrak{a}$, the corresponding point $\mathbf{x}$ is undefined, and $\mathfrak{a}$ is called a **base point**.

7 Recall the generalization in Section 10.7 of barycentric coordinates in the plane. The outer vertices of the quintangle are base points.

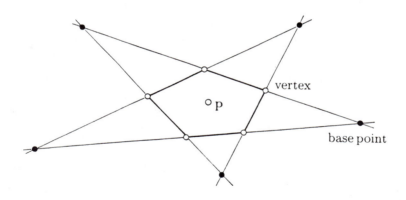

Figure 27.9: Base points in a quintangle.

8 Consider a rational Bézier curve of degree n given by

$$\varkappa(t) = \sum_{i=0}^{n} \mathbb{b}_i B_i^n(t)$$

where $\mathbb{b}_0 = \mathfrak{o}$. Then $t = 0$ is a base point. Moreover, one can factor out t. Hence the curve is of degree $n - 1$ only, and it starts at $\mathbb{b}_1$, i.e., $\varkappa(0) = \mathbb{b}_1$.

9 Consider a rational triangular Bézier surface of degree n given by

$$\varkappa(u, v, w) = \sum \mathbb{b}_{ijk} B^n_{ijk}(u, v, w)$$

where $\mathbb{b}_{000} = \mathbb{o}$. Then $\mathbf{u}_0 = [0\ 0\ 1]^t$ is a base point: For $\mathbf{u} \to \mathbf{u}_0$ with $u/v = \alpha/\beta$ fixed one has

$$\varkappa(u, v, w) \to \mathbb{b}_{10(n-1)}\alpha + \mathbb{b}_{01(n-1)}\beta$$

which is a point on the line $\mathbb{b}_{10(n-1)}\mathbb{b}_{01(n-1)}$.

28 Some Elimination Methods

Implicitization, inversion, and the intersection of curves and surfaces are closely related to the so-called elimination theory which was intensively discussed in algebraic geometry about 150 years ago. Elimination methods provide criteria for the existence of common roots of several polynomials, in particular for the common roots of two univariate polynomials and the common roots of three bivariate polynomials.

Literature: Gröbner, Salmon, van der Waerden

28.1 Sylvester's Method

If one wants to know whether some given polynomials have a common root without computing their roots explicitly, one can employ a resultant. A **resultant** is an expression in the coefficients of several polynomials which vanishes if all the polynomials have a **common root**.

A simple method for constructing a resultant of two polynomials in one variable is due to **Sylvester**. Let

$$a(x) = a_0 + a_1 x + \cdots + a_m x^m = \mathfrak{a}^t \varkappa_m \ ,$$
$$b(x) = b_0 + b_1 x + \cdots + b_n x^n \ \ = \mathfrak{b}^t \varkappa_n \ ,$$

where $a_m \neq 0$, $b_n \neq 0$ and $m \geq 1, n \geq 1$.

Multiplying $a(x)$ by $1, x, \ldots, x^{n-1}$ and $b(x)$ by $1, x, \ldots, x^{m-1}$ gives $m+n$ polynomials. Equating these polynomials with zero establishes the system

$$
\begin{array}{c} n \\ \\ m \\ \\ \end{array}
\begin{bmatrix}
a_0 & \cdots & a_m & & & \\
 & \ddots & & \ddots & & \\
 & & a_0 & \cdots & a_m \\
b_0 & \cdots & b_n & & & \\
 & \ddots & & \ddots & \\
 & & b_0 & \cdots & b_n
\end{bmatrix}
\begin{bmatrix}
1 \\ \vdots \\ x^{n-1} \\ x^n \\ \vdots \\ x^p
\end{bmatrix}
=
\begin{bmatrix}
0 \\ \vdots \\ 0 \\ 0 \\ \vdots \\ 0
\end{bmatrix}
$$

or $S\varkappa_p = \mathrm{o}$ where $p = n + m - 1$.

Obviously, the normal curve $\varkappa_p$ at a common root of $a(x)$ and $b(x)$ solves the linear system $Sy = \mathrm{o}$. Thus, if $\det S \neq 0$, there is no common root for $a(x)$ and $b(x)$. The converse is also true. Namely let $x_1, \ldots, x_m$ be the real and complex roots of $a(x)$, and let

$$
\mathbb{p}_i = \frac{d^r}{dx^r} \varkappa_p \Big|_{x = x_i}
$$

where x_i occurs r times in $x_1, \ldots, x_{i-1}$. Since $[a_0 \ldots a_m 0 \ldots 0]\mathbb{p}_i = a^{(r)}(x_i) = 0$, any $\mathbb{p}_i$ solves the first n equations of $S\varkappa = \mathrm{o}$.

Since multiples of $a(x)$ are the only polynomials of degree m having the roots $x_1, \ldots, x_m$, the linear system for $c_0, \ldots, c_m$,

$$
[\mathbb{p}_1 \ldots \mathbb{p}_m]^{\mathrm{t}}[c_0 \ldots c_m 0 \ldots 0]^{\mathrm{t}} = \mathrm{o} ,
$$

has a one-dimensional solution. Thus $\mathbb{p}_1, \ldots, \mathbb{p}_m$ span the solution of the first n equations of $Sy = \mathrm{o}$.

Similarly, there are columns $\mathbb{p}_{m+1}, \ldots, \mathbb{p}_{m+n}$ constructed from the roots $x_{m+1}, \ldots, x_{m+n}$ of $b(x)$. If $a(x)$ and $b(x)$ have no common root, then $\mathbb{p}_1, \ldots, \mathbb{p}_{m+n}$ are independent and o is the only column solving all equations of $Sy = \mathrm{o}$, i.e., $\det S \neq 0$. Thus, $\det S$ is a resultant of both polynomials.

Example 1: Consider the two polynomials

$$
a(x) = 1 + 1x - 1x^2 - 1x^3 ,
$$
$$
b(x) = 0 + 2x - 1x^2 - 2x^3 + 1x^4
$$

with common roots $x = +1$ and $x = -1$. Sylvester's method gives the linear system

$$\begin{bmatrix} 1 & 1 & -1 & -1 & & & \\ & 1 & 1 & -1 & -1 & & \\ & & 1 & 1 & -1 & -1 & \\ & & & 1 & 1 & -1 & -1 \\ 0 & 2 & -1 & -2 & 1 & & \\ & 0 & 2 & -1 & -2 & 1 & \\ & & 0 & 2 & -1 & -2 & 1 \end{bmatrix} y = o \ .$$

The solution $y = [1\ 1\ 1\ 1\ 1\ 1\ 1]^t + \lambda[0\ 1\ 0\ 1\ 0\ 1\ 0]^t$ can be computed by the Gauss–Jordan algorithm. It intersects the normal curve x_p for $\lambda = 0$ and $\lambda = -2$.

28.2 Cayley's Method

A resultant based on a symmetric $n \times n$ matrix only where $m \leq n$ is due to **Bezout**. A nice derivation of it was given by **Cayley**: Let the two polynomials from above be combined into

$$\mathbf{a}(t) = \mathbf{a}_0 + \mathbf{a}_1 x + \cdots + \mathbf{a}_n x^n,$$

where $\mathbf{a}_i = [a_i \quad b_i]^t$ and $\mathbf{a}_n \neq o$. The expression

$$d = \frac{\det[\mathbf{a}(x) \quad \mathbf{a}(s)]}{x - s}$$

has the following properties. Its numerator is a polynomial in x and s which vanishes if $x = s$. Hence, it is divisible by $x - s$. As a consequence, d is a polynomial of degree $n - 1$ in x and s and it can be written as

$$d = [1 \quad s \ \cdots \ s^p] \begin{bmatrix} c_{00} & \cdots & c_{0p} \\ \vdots & C & \vdots \\ c_{p0} & \cdots & c_{pp} \end{bmatrix} \begin{bmatrix} 1 \\ x \\ \vdots \\ x^p \end{bmatrix}$$

where $p = n - 1$. If $x = x_0$ is a common root, then $\mathbf{a}(x_0) = o$ and $d = 0$ for all s. Conversely, if $d = 0$ for all s, the determinant $\det[\mathbf{a}(x_0) \quad \mathbf{a}(s)]$

equals zero. Since $\mathbf{a}(x_0)$ and $\mathbf{a}(s)$ cannot be linearly dependent for all s unless $a(x)$ is a multiple of $b(x)$, one has $\mathbf{a}(x_0) = 0$, i.e., x_0 is a common root. Moreover, $d = 0$ for all s and $x = x_0$ if and only if

$$
\begin{bmatrix} c_{00} & \cdots & c_{0p} \\ \vdots & C & \vdots \\ c_{p0} & \cdots & c_{pp} \end{bmatrix}
\begin{bmatrix} 1 \\ x_0 \\ \vdots \\ x_0^p \end{bmatrix}
=
\begin{bmatrix} 0 \\ 0 \\ \vdots \\ 0 \end{bmatrix}.
$$

This homogeneous linear system is solvable only if $\det C = 0$. Thus, $\det C$ is a resultant for both polynomials.

28.3 Computing Cayley's Matrix

Some algebraic manipulations show that the entries of **Cayley's matrix** C can be computed by using

$$
c_{ij} = \sum_k \det \begin{bmatrix} \mathbf{a}_k & \mathbf{a}_{i+j+1-k} \end{bmatrix},
$$

where k runs from $max\{0, i + j + 1 - n\}$ to $min\{i, j\}$. Note that C is symmetric.

Example 2: Cayley's matrix can be written as the sum of n simple matrices, e.g., for $n = 4$:

$$
C = \begin{bmatrix} 01 & & & \\ & & & \\ & & & \\ & & & \end{bmatrix}
+ \begin{bmatrix} & 02 & & \\ 02 & 12 & & \\ & & & \\ & & & \end{bmatrix}
$$

$$
+ \begin{bmatrix} & & 03 & \\ & 03 & 13 & \\ 03 & 13 & 23 & \\ & & & \end{bmatrix}
+ \begin{bmatrix} & & & 04 \\ & & 04 & 14 \\ & 04 & 14 & 24 \\ 04 & 14 & 24 & 34 \end{bmatrix},
$$

where ij stands for $\det \begin{bmatrix} \mathbf{a}_i & \mathbf{a}_j \end{bmatrix}$. Note that $x = 0$ is a common root if $\mathbf{a}_0 = \mathbf{o}$.

Example 3: If the polynomials a and b are as in Example 1, Cayley's method yields the linear system

$$\begin{bmatrix} 2 & -1 & -2 & 1 \\ -1 & -1 & 1 & 1 \\ -2 & 1 & 2 & -1 \\ 1 & 1 & -1 & -1 \end{bmatrix} \begin{bmatrix} 1 \\ x_1 \\ x_2 \\ x_3 \end{bmatrix} = \mathbf{o} .$$

Its solution, $x = \begin{bmatrix} 1 & 1 & 1 & 1 \end{bmatrix}^t + \lambda \begin{bmatrix} 0 & 1 & 0 & 1 \end{bmatrix}^t$, intersects the normal curve x_3 for $\lambda = 0$ and $\lambda = -2$.

28.4 Dixon's Method

Dixon generalized Cayley's method to three linearly independent polynomials in two variables. Let the three linearly independent polynomials be the coordinates of

$$\mathbf{a}(x, y) = \mathbf{a}_{0,0} + \mathbf{a}_{1,0}x + \mathbf{a}_{0,1}y + \cdots + \mathbf{a}_{n,m}x^n y^m,$$

where $\mathbf{a}_{i,j} = \begin{bmatrix} a_{i,j} & b_{i,j} & c_{i,j} \end{bmatrix}^t$ and $\mathbf{a}_{n,m} \neq \mathbf{o}$. The expression

$$f(x, y, s, t) = \frac{\det \begin{bmatrix} \mathbf{a}(x, y) & \mathbf{a}(x, t) & \mathbf{a}(s, t) \end{bmatrix}}{(x - s)(y - t)}$$

has the following properties. Its numerator is a polynomial in x, y, s, t which vanishes if $x = s$ or $y = t$. Hence, it is divisible by $x - s$ and $y - t$. As a consequence, f is a polynomial of

degree $2n - 1, m - 1, n - 1, 2m - 1$ in x, y, s and t, respectively.

It can be written as $f = \sum d_{\alpha\beta\gamma\delta} x^\alpha y^\beta s^\gamma t^\delta$ or in matrix notation as

$$f = \begin{bmatrix} \ldots s^\gamma t^\delta \ldots \end{bmatrix} \begin{bmatrix} d_{00} & \cdots & d_{0r} \\ \vdots & D & \vdots \\ d_{r0} & \cdots & d_{rr} \end{bmatrix} \begin{bmatrix} \vdots \\ x^\alpha y^\beta \\ \vdots \end{bmatrix}$$

where $r = 2mn - 1$.

If $(x,y) = (x_0, y_0)$ is a common root, then $\mathbf{a}(x_0, y_0) = 0$ and therefore $f = 0$ for all s and all t, i.e.,

$$
\begin{bmatrix} d_{00} & \cdots & d_{0r} \\ \vdots & D & \vdots \\ d_{r0} & \cdots & d_{rr} \end{bmatrix} \begin{bmatrix} \vdots \\ x_0^\alpha y_0^\beta \\ \vdots \end{bmatrix} = \begin{bmatrix} 0 \\ \vdots \\ 0 \end{bmatrix} .
$$

This homogeneous linear system is solvable if and only if $\det D = 0$.

Conversely, if $\det D = 0$, one has $f = 0$ for all s and t and some fixed $(x,y) = (x_0, y_0)$. Since $\mathbf{a}(x_0, y_0), \mathbf{a}(x_0, t)$, and $\mathbf{a}(s, t)$ define a point, curve, and surface, respectively, these coordinate columns are linearly dependent only if $\mathbf{a}(x_0, y_0) = \mathbf{o}$. Hence (x_0, y_0) is a common root. Thus, $\det D$ is a resultant for the three polynomials.

Note that this resultant is of degree $2mn$ in each coefficient of the given polynomials.

28.5 Computing Dixon's Matrix

Some algebraic calculations show that the coefficients of the polynomial $f = f(x, y, s, t)$ can be computed by

$$
d_{\alpha\beta\gamma\delta} = \sum \det \begin{bmatrix} \mathbf{a}_{r,s} & \mathbf{a}_{r',s'+\beta+1} & \mathbf{a}_{r''+\gamma+1,s''} \end{bmatrix}
$$

for all solutions of $r + r' + r'' = \alpha$ and $s + s' + s'' = \delta$.

The position of $d_{\alpha\beta\gamma\delta}$ in the matrix depends on the ordering of $x^\alpha y^\beta$ and $s^\gamma t^\delta$. One of the possible address functions is

$$
i = 2m\gamma + \delta , \qquad j = \alpha m + \beta .
$$

Example 4: In the case where $n = 2$ and $m = 1$, one has

$$
\mathbf{a}(x, y) = \begin{bmatrix} 1 & x & x^2 \end{bmatrix} \begin{bmatrix} \mathbf{a}_{00} & \mathbf{a}_{01} \\ \mathbf{a}_{10} & \mathbf{a}_{11} \\ \mathbf{a}_{20} & \mathbf{a}_{21} \end{bmatrix} \begin{bmatrix} 1 \\ y \end{bmatrix} .
$$

Dixon's method gives

$$
f = \begin{bmatrix} 1 & t & s & st \end{bmatrix}
\begin{bmatrix}
\begin{array}{ll} 00.01.10 \\ \end{array} & \begin{array}{l} 00.01.20 \\ +00.11.10 \end{array} & \begin{array}{l} 00.21.10 \\ +00.11.20 \end{array} & 00.21.20 \\[2em]
\begin{array}{l} 00.01.11 \end{array} & \begin{array}{l} 00.01.21 \\ +01.11.10 \end{array} & \begin{array}{l} 01.11.20 \\ +01.21.10 \end{array} & 01.21.20 \\[2em]
\begin{array}{l} 00.01.20 \end{array} & \begin{array}{l} 00.11.20 \\ +10.01.20 \end{array} & \begin{array}{l} 00.21.20 \\ +10.11.20 \end{array} & 10.21.20 \\[2em]
\begin{array}{l} 00.01.21 \end{array} & \begin{array}{l} 00.11.21 \\ +10.01.21 \end{array} & \begin{array}{l} 01.21.20 \\ +10.11.21 \end{array} & 11.21.20
\end{array}
\end{bmatrix}
\begin{bmatrix} 1 \\ x \\ x^2 \\ x^3 \end{bmatrix},
$$

where $ij.kl.pq$ stands for $\det [\mathbf{a}_{ij} \quad \mathbf{a}_{kl} \quad \mathbf{a}_{pq}]$.

Remark 1: Some special cases can occur when using the above formulas to compute the entries of Dixon's matrix. Namely, indices may be out of range, determinants may be zero because of two equal columns, and pairs of determinants may sum to zero since two columns are interchanged.

28.6 Triangular Matrices

Dixon's method above fails if $\mathbf{a}(x, y)$ is of total degree, i.e., if $\mathbf{a}_{i,k} = \mathbf{o}$ for $i + k > n$. Then one has $d_{\alpha\beta\gamma\delta} = 0$ for $\alpha + \beta > 2n - 1$ and for $\gamma + \delta > 2n-1$. Furthermore, one can show that $d_{\alpha\beta\gamma\delta} = 0$ for $\gamma+\delta > n-1$. As a consequence, the number of effective pairs $s^\gamma t^\delta$ and $x^\alpha y^\beta$ reduces to $p = \frac{1}{2}n(n+1)$ and $q = 2n^2 - n$, respectively, while the matrix D becomes a $p \times q$ matrix D_1 which is wider than it is high. One has

$$
\begin{bmatrix} \ldots s^\gamma t^\delta \ldots \end{bmatrix}
\begin{bmatrix}
d_{0,0} & q\cdots & d_{0,q-1} \\
\vdots & & \vdots \\
d_{p-1,0} & \cdots & d_{p-1,q-1}
\end{bmatrix}
\begin{bmatrix} \vdots \\ x^\alpha y^\beta \\ \vdots \end{bmatrix} = 0
$$

for all s and t, which results in p linear equations

$$
\begin{bmatrix}
d_{0,0} & \cdots & d_{0,q-1} \\
\vdots & D_1 & \vdots \\
d_{p-1,0} & \cdots & d_{p-1,q-1}
\end{bmatrix}
\begin{bmatrix} \vdots \\ x^\alpha y^\beta \\ \vdots \end{bmatrix} =
\begin{bmatrix} 0 \\ \vdots \\ 0 \end{bmatrix}.
$$

Following Sylvester's ideas, Dixon's method uses the $\frac{1}{2}n(n-1)$ factors $x^\varrho y^\sigma$, $\varrho + \sigma \le n - 1$ to augment the matrix, with rows representing the three polynomial coordinates of $\mathbf{a}(x,y)$. This results in $q - p = \frac{3}{2}n(n-1)$ further linear equations

$$
\begin{bmatrix}
d_{p,0} & \cdots & d_{p,q-1} \\
\vdots & D_2 & \vdots \\
d_{q-1,0} & \cdots & d_{q-1,q-1}
\end{bmatrix}
\begin{bmatrix}
\vdots \\
x^\varrho y^\sigma \\
\vdots
\end{bmatrix}
=
\begin{bmatrix}
0 \\
\vdots \\
0
\end{bmatrix} .
$$

Both systems lead to q homogeneous equations for the q unknowns, $x^\alpha y^\beta$. The system is solvable only if its determinant vanishes. It is a resultant again.

Example 5: If the total degree equals $n = 2$, one has

$$
\mathbf{a}(x,y) = \begin{bmatrix} 1 & x & x^2 \end{bmatrix}
\begin{bmatrix}
\mathbf{a}_{00} & \mathbf{a}_{01} & \mathbf{a}_{02} \\
\mathbf{a}_{10} & \mathbf{a}_{11} \\
\mathbf{a}_{20}
\end{bmatrix}
\begin{bmatrix}
1 \\
y \\
y^2
\end{bmatrix} , \quad
\mathbf{a}_{ik} =
\begin{bmatrix}
a_{ik} \\
b_{ik} \\
c_{ik}
\end{bmatrix} .
$$

Dixon's combined method yields the two systems

$$
\begin{bmatrix}
00.00.10 & 00.02.10 & 0 & \begin{matrix} 00.01.20 \\ + \ 00.11.10 \end{matrix} & 00.02.20 & 00.11.20 \\
\begin{matrix} 00.01.11 \\ +00.02.10 \end{matrix} & \begin{matrix} 00.02.10 \\ +01.02.10 \end{matrix} & 0 & \begin{matrix} 00.02.20 \\ +01.11.10 \end{matrix} & 01.02.20 & 01.11.20 \\
00.01.20 & 00.02.20 & 0 & \begin{matrix} 00.11.20 \\ +10.01.20 \end{matrix} & 10.02.20 & 01.11.20
\end{bmatrix}
\begin{bmatrix}
1 \\
y \\
y^2 \\
x \\
xy \\
x^2
\end{bmatrix}
=
\begin{bmatrix}
0 \\
0 \\
0
\end{bmatrix}
$$

and

$$
\begin{bmatrix}
a_{00} & a_{01} & a_{02} & a_{10} & a_{11} & a_{20} \\
b_{00} & b_{01} & b_{02} & b_{10} & b_{11} & b_{20} \\
c_{00} & c_{01} & c_{02} & c_{10} & c_{11} & c_{20}
\end{bmatrix}
\begin{bmatrix}
1 \\
y \\
y^2 \\
x \\
xy \\
x^2
\end{bmatrix}
=
\begin{bmatrix}
0 \\
0 \\
0
\end{bmatrix} .
$$

Remark 2: It should be pointed out that the vanishing of the resultant is a necessary but not always sufficient condition for the existence of a common root.

28.7 Notes and Problems

1 Sylvester's dialytic method, published in 1840, involved the elimination of n polynomials in $n-1$ variables.

2 The resultant of n polynomials in $n-1$ variables coincides with the resultant of the corresponding n polynomials in n homogeneous variables.

3 The resultant of two polynomials of degree n and m in one variable is of degree m in the coefficients of the polynomial of degree n.

4 Given three polynomials of total degree n in two variables, the resultant is of degree n^2 in the coefficients of each of the three polynomials.

5 Given three polynomials of degree n, m in two variables, the resultant is of degree $2nm$ in the coefficients of each of the three polynomials.

6 If some polynomials have a common root, then any linear combination has the same root.

7 A common proof of **Bezout's theorem** is as follows. Let two curves of the projective plane in general position be given by the equations

$$f(x, y, z) = f_0 + f_1 z + \cdots + f_m z^m ,$$
$$g(x, y, z) = g_0 + g_1 z + \cdots + g_n z^n ,$$

where $f_i = f_i(x, y)$ and $g_j = g_j(x, y)$ are homogeneous polynomials of degree $m - i$ and $n - j$ in x and y. For any fixed x, y both f and g are polynomials of degree m and n in z with a common root if the line $\mathcal{L}$ defined by $x = \lambda y$ meets an intersection point of both curves. On the other hand, the resultant of f and g is a polynomial of degree $m \cdot n$ in λ, cf. Note 3, which has $m \cdot n$ roots. This proves the theorem.

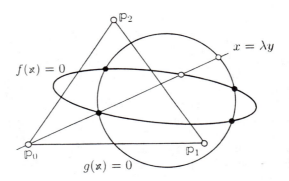

Figure 28.1: Proof of Bezout's theorem.

8 The proof given in the previous Note also demonstrates how to find the intersection points of two implicit algebraic curves, also see Section 29.6.

29 Implicitization, Inversion and Intersection

Not every algebraic curve and surface has a rational parametrization. On the other hand, one can represent every parametric curve or surface implicitly by a system of polynomial equations. The process of obtaining such equations is called implicitization. It is closely related to the elimination theory discussed in the previous chapter. Two other methods related to the elimination methods are inversion methods, which procure the parameter value of a given point on a parametric curve or surface, and methods for computing the intersections of algebraic curves and surfaces.

Literature: Brieskorn·Knörrer, Hoffmann, Salmon

29.1 Parametric Curves in the Plane

A polynomial curve of degree n in the affine plane given by

$$\mathbf{x}(t) = \mathbf{a}_0 + \mathbf{a}_1 t + \cdots + \mathbf{a}_n t^n , \qquad \mathbf{a}_i = [a_i \quad b_i]^t ,$$

can be represented by the equation

$$\mathbf{p}(t) = \mathbf{p}(\mathbf{x}, t) = [\mathbf{a}_0 - \mathbf{x}] + \mathbf{a}_1 t + \cdots + \mathbf{a}_n t^n = \mathbf{o}$$

or, at full length, by

$$p(t) = (a_0 - x) + a_1 t + \cdots + a_n t^n = 0$$
$$q(t) = (b_0 - y) + b_1 t + \cdots + b_n t^n = 0 .$$

A point $\mathbf{x}$ lies on the curve if and only if both components of $\mathbf{q}(\mathbf{x}, t)$ have a common root. Hence, if $C = C(\mathbf{x})$ is a matrix such that $\det C = 0$ is a resultant for $p(t)$ and $q(t)$, then

$$f(\mathbf{x}) = \det C = 0$$

is an implicit equation of the curve $\mathbf{x}(t)$.

A plane rational curve of degree n,

$$\mathbf{x}(t) = \frac{\mathbf{a}_0 + \mathbf{a}_1 t + \cdots + \mathbf{a}_n t^n}{\alpha_0 + \alpha_1 t + \cdots + \alpha_n t^n} \ , \qquad \mathbf{a}_i = [a_1 \quad b_i]^t \ ,$$

can be represented by the equation

$$\mathbf{q}(t) = [\mathbf{a}_0 - \mathbf{x}\alpha_0] + [\mathbf{a}_1 - \mathbf{x}\alpha_1] \, t + \cdots + [\mathbf{a}_n - \mathbf{x}\alpha_n] \, t^n = \mathbf{o} \ .$$

A point $\mathbf{x}$ lies on the curve if and only if both components of $\mathbf{q}(\mathbf{x}, t)$ have a common root. Hence if C is a matrix whose determinant is a resultant for the components of $\mathbf{q}(t)$, then

$$f(\mathbf{x}) = \det C = 0$$

is an implicit equation of the curve $\mathbf{q}(t)$.

Furthermore, if C is Sylvester's or Cayley's matrix the **inversion problem** can be tackled by solving the homogeneous linear system $C[1 \quad t \ldots t^p]^t = \mathbf{o}$. One can apply Cramer's rule to get

$$t = -\det C_{01}/\det C_{00} \ ,$$

where C_{ij} is obtained from C by taking out the ith row and jth column.

Remark 1: When using Bezout's resultant the elements of C are sums of the determinants

$$ik = \det [\mathbf{a}_i - \mathbf{x}\alpha_i \quad \mathbf{a}_k - \mathbf{x}\alpha_k]$$

which can be written as

$$ik = \det \begin{bmatrix} \mathbf{x} & \mathbf{a}_i & \mathbf{a}_k \\ 1 & \alpha_i & \alpha_k \end{bmatrix} = \mathbf{u}^t \mathbf{x} + u_0 = \mathsf{u}_{ik}^t \mathbf{x} \ .$$

Hence, each entry of the matrix C is linear in x and y. Therefore Bezout's resultant is a polynomial of degree n in x and y. The same holds true for Sylvester's resultant for integral curves.

Remark 2: The implicit equation of an integral curve of degree n can be written as

$$f(\mathbf{x}) = \det{}^{n} [\mathbf{x} \; \mathbf{a}_n] + \text{(terms of degree } < n) = 0 .$$

The reason is that only $0n = \det [\mathbf{a}_0 - \mathbf{x}, \mathbf{a}_n]$ contributes to terms of order n (see Section 28.3). As a geometric consequence, each parametric integral curve of degree n has an n-fold intersection with the ideal line.

Example 1: For the integral cubic

$$x = 1 + t - t^2 - t^3 = -(t+1)^2(t-1) ,$$
$$y = 1 - t - t^2 + t^3 = +(t+1)(t-1)^2$$

one gets

$$01 = x + y - z, \;\; 02 = x - y, \;\; 12 = -2, \;\; 03 = 2 - x - y, \;\; 13 = 0, \;\; 23 = -2$$

which has the implicit equation

$$f = \det \begin{bmatrix} 01 & 02 & 03 \\ 02 & 12+03 & 13 \\ 03 & 13 & 23 \end{bmatrix} = -(x+y)^3 + 8xy = 0 .$$

Note that $\det{}^3 [\mathbf{x} \;\; \mathbf{a}_n] = -(x+y)^3$ is the leading term.

Furthermore, for any point $[x \;\; y]^t$ on the curve, one gets

$$-\det \begin{bmatrix} 02 & 13 \\ 03 & 23 \end{bmatrix} \Big/ \det \begin{bmatrix} 12+03 & 13 \\ 13 & 23 \end{bmatrix} = \frac{x-y}{y+x} = t .$$

For example, the point $[1 \;\; 1]^t$ is parameterized by $t = 0$.

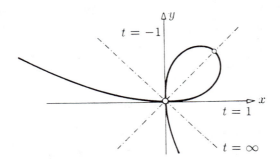

Figure 29.1: Example 1, $8xy = (x+y)^3$.

29.2 Parametric Space Curves

Consider a parametric curve in $\mathcal{A}^3$

$$x = a(t) , \qquad y = b(t) , \qquad z = c(t) .$$

Then $x = s$, $y = b(t)$, $z = c(t)$ represent a space cylinder containing the curve. One can obtain the implicit equation of this cylinder $f(\mathbf{x}) = 0$ by eliminating t in $y = b(t)$ and $z = c(t)$. Similarly, one gets the equations of two further cylinders by replacing $b(t)$ or $c(t)$ by the parameter s. In general, the common solution of two such equations describes a curve in space.

If the curve is a rational or integral polynomial curve of degree n, the equations of the three cylinders are also of degree n.

Given a point $\mathbf{x}$ on the curve one can solve the inversion problem by considering some suitable projection of the point and curve, e.g., onto the xy-plane.

Notice that the three cylinders depend on the affine system. There is a cylinder of degree n in any direction of $\mathcal{A}^3$.

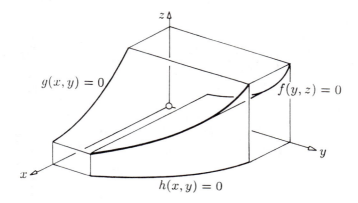

Figure 29.2: Cylinders, projecting a space curve.

29.3 Normal Curves

Implicitization and inversion are rather simple and do not need a resultant if the parametric curve spans $\mathcal{A}^n$, where n is the degree of the curve. In homogeneous coordinates such a curve takes on the form

$$\varkappa \varrho = \mathbb{P} \mathfrak{k}_n .$$

The matrix $\mathbb{P}$ is invertible since the curve spans $\mathcal{P}^n$. Hence, $\varkappa$ lies on the curve $\mathbb{P}\mathfrak{k}_n$ if

$$y \sigma = \mathbb{P}^{-1} \varkappa$$

lies on the normal curve $\mathfrak{k}_n$. Thus, the task of finding an implicit representation for $\varkappa(t)$ can be reduced to the implicitization of the normal curve.

The normal curve lies on the **normal quadrics**

$$y_0 y_2 = y_1^2 , \qquad y_0 y_3 = y_1 y_2 , \qquad \cdots , \qquad y_{n-2} y_n = y_{n-1}^2 ,$$

and also on the cubic surfaces $y_0^2 y_3 = y_1^3$, etc. Sufficiently many of these equations define the normal curve. Hence, by substituting some multiple of

$\mathbb{P}^{-1}x$ for y in one of these equations, one obtains an implicit representation of x, as illustrated in Figure 29.3.

The inversion problem also has a simple solution. Namely, the parameter of some point $x(t)$ is given by $t = y_1/y_0$.

Example 2: If $n = 2$, the curve $x(t)$ is a conic section while the normal curve has the equation $y_0 y_2 = y_1^2$. For example, if

$$x = [1 + t^2 \quad 1 - t^2 \quad 2t]^t$$

then

$$y = [x_0 + x_1 \quad x_2 \quad x_0 - x_2]^t$$

and

$$y_0 y_2 - y_1^2 = (x_0 + x_1)(x_0 - x_1) - x_2^2 = x_1^2 + x_2^2 - x_0^2$$

which represents the unit circle for $x_0 = 1$. Furthermore, the point $x = [1 \quad 0.8 \quad 0.6]^t$ is parameterized by

$$t = \frac{x_2}{x_0 + x_1} = \frac{0.6}{1.8} = \frac{1}{3} .$$

Example 3: If $n = 3$, the curve $x(t)$ is a space cubic. The normal curve y is the intersection of the parabolic cylinder $y_0 y_2 = y_1^2$, the hyperbolic paraboloid $y_0 y_3 = y_1 y_2$, the cone $4y_1 y_3 = y_2^2$, and also the cubic cylinder $y_0 y_3^2 = y_2^3$, as illustrated in Figure 29.3.

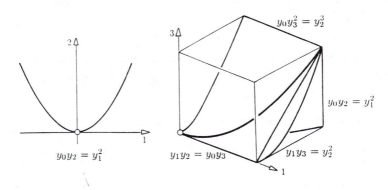

Figure 29.3: Normal curves and equations.

Example 4: Consider the curve in affine space

$$\mathbf{x} = \begin{bmatrix} 0 \\ 1 \\ 0 \end{bmatrix} + \begin{bmatrix} 2 & -1 & 0 \\ 0 & 1 & 0 \\ 0 & 3 & -2 \end{bmatrix} \begin{bmatrix} t \\ t^2 \\ t^3 \end{bmatrix} .$$

The normal curve is given by

$$\mathbf{y} = [- \begin{bmatrix} 1 \\ 2 \\ 3 \end{bmatrix} + \begin{bmatrix} 1 & 1 & 0 \\ 0 & 2 & 0 \\ 0 & 3 & -1 \end{bmatrix} \begin{bmatrix} x_1 \\ x_2 \\ x_3 \end{bmatrix}]\frac{1}{2}$$

which implies, e.g., that the curve lies on the parabolic cylinder

$$4(y_2^2 - y_1) = (x_1 + x_2 - 1)^2 - 4(x_2 - 1) = 0 .$$

Note, that in this example $x_0 = y_0 = 1$. Furthermore, the point $\mathbf{x} = [1 \quad 2 \quad 1]^t$ has the parameter

$$t = y_1 = x_2 - 1 = 1 .$$

29.4 Parametric Tensor Product Surfaces

A polynomial surface of degree m, n in the affine space given by

$$\mathbf{y}(\mathbf{u}) = \mathbf{a}_{00} + \mathbf{a}_{10}u + \mathbf{a}_{01}v + \cdots + \mathbf{a}_{mn}u^m v^n$$

$$\mathbf{a}_{ij} = [a_{ij} \quad b_{ij} \quad c_{ij}]^t$$

can be represented by the equation

$$\mathbf{p}(\mathbf{u}) = [\mathbf{a}_{0,0} - \mathbf{y}] + \mathbf{a}_{1,0}u + \mathbf{a}_{0,1}v + \cdots + \mathbf{a}_{m,n}u^m v^n = \mathbf{o} .$$

A point $\mathbf{y}$ lies on the surface if and only if all three components of $\mathbf{p}(\mathbf{u})$ have a common root. Hence, if D denotes Dixon's matrix computed from the components of $\mathbf{p}$, one has that

$$F(\mathbf{y}) = \det D = 0$$

is an implicit equation of the surface $\mathbf{p}(u, v)$.

Furthermore, the parameter values of some point $\mathbf{y}$ on this surface can be determined by solving the corresponding homogeneous linear system.

A rational surface of degree m, n,

$$\mathbf{y}(u, v) = \frac{\mathbf{a}_{0,0} + \cdots + \mathbf{a}_{m,n} u^m v^n}{\alpha_{0,0} + \cdots + \alpha_{m,n} u^m v^n} ,$$

can be represented by the equation

$$\mathbf{q}(u, v) = [\mathbf{a}_{0,0} - \mathbf{y} \alpha_{0,0}] + \cdots + [\mathbf{a}_{m,n} - \mathbf{y} \alpha_{m,n}] u^m v^n = \mathbf{o} .$$

Similarly, as before,
$$F(\mathbf{y}) = \det D = 0$$

is an implicit equation of the surface, where D denotes Dixon's matrix obtained from $\mathbf{q}$.

Furthermore, the parameter values of some point $\mathbf{y}$ on this surface can be determined by solving the corresponding homogeneous linear system. Using Cramer's rule one gets

$$u = -\det D_{01}/\det D_{00} , \quad v = \det D_{02}/\det D_{00} ,$$

where D_{ij} is obtained from D by taking out the ith row and jth column.

Remark 3: The entries of D can be written as

$$d_{i,j} = \mathbf{u}_{i,j}^{\mathrm{t}} \mathbf{y} + u_{i,j} ,$$

i.e., they are linear in y_1, y_2, and y_3. Hence, the degree of the resultant is less than or equal to $2mn$.

Remark 4: The order of a tensor product surface $\mathbf{y}(\mathbf{u})$ of degree m, n can be tracked down as follows: Consider the intersection of the surface with two arbitrary planes $a(\mathbf{y}) = 0$ and $b(\mathbf{y}) = 0$. Both intersection curves correspond to curves $a(\mathbf{y}(\mathbf{u})) = 0$ and $b(\mathbf{y}(\mathbf{u})) = 0$ of degree $m+n$ in the $\mathbf{u}$-plane which intersect in $(m+n)^2$ points because of Bezout's theorem. Both parameter curves have m and n points at the ideal points of the u- and v-axes, respectively, and thus m^2 and n^2 common points there. Therefore, both curves intersect in only $2mn$ further points. This corresponds to the fact that the order of the tensor product surface $\mathbf{y}$ equals $2mn$, see Note 5.

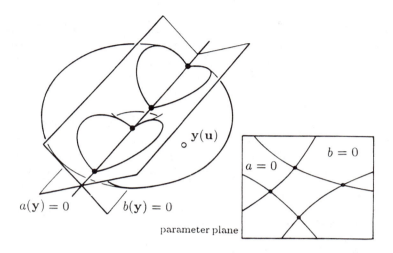

Figure 29.4: Order of a parametric surface.

Remark 5: A point in the projective extension of the **u**-plane such that $y(u) = o$ is called a **base point**. Every base point of y reduces the order of $y(u)$ by one.

29.5 Parametric Triangular Surfaces

If the coefficients $\mathbf{a}_{i,j}$ of a polynomial surface are zero for $i + j > n$, their matrix is triangular. Therefore such a surface is called a **triangular surface** or **surface of total degree** n. Such a surface has an appealing property: Any straight line of the **u**-plane corresponds to a curve of degree n on the surface. The implicitization and inversion problem can be solved as above by using a suitable resultant. Although the order of a tensor product surface of order n, n is $\leq 2n^2$, a parametric triangular surface of total degree n is only of order $\leq n^2$. In order to verify this consider two arbitrary planes given by $a(\mathbf{x}) = 0$ and $b(\mathbf{x}) = 0$. Their intersection with the surface gives two curves $a(\mathbf{x}(\mathbf{u}))$ and $b(\mathbf{x}(\mathbf{v}))$ in the **u**-plane. Bezout's theorem implies that both curves intersect in n^2 points. These n^2 points corresponds to n^2 points on the surface $\mathbf{y}$ and form the intersection of a line with the surface. Hence, the surface is of order n^2 only.

Example 5: A rational quadratic triangular surface $\mathbf{y}(\mathbf{u})$ is of order 4 in general. However, it represents a non-degenerate quadric if there are two base points. Every intersection with a plane $a(\mathbf{y}) = 0$ corresponds to a conic section $a(\mathbf{y}(\mathbf{u})) = 0$ in the $\mathbf{u}$-plane containing both base points. Moreover, all isolines meet in a point of the quadric.

A sphere parameterized via a stereographic projection provides an example of a triangular quadratic representation of a quadric. All plane intersections correspond to circles in the $\mathbf{u}$-plane. The (imaginary) circular points are the two base points which reduce the order of the surface from 4 to 2.

29.6 Intersections

The methods discussed in this chapter can be applied so as to compute the intersection of two implicit algebraic curves or surfaces.

Curve-Curve intersection: Consider two planar algebraic curves given implicitly by $f(\mathbf{x}) = 0$ and $g(\mathbf{x}) = 0$, where

$$f = f_0 + f_1 x + \cdots + f_m x^m , \qquad g = g_0 + g_1 x + \cdots + g_n x^n ,$$

and the f_i and g_j are polynomials in y of degrees $m - i$ and $n - j$, respectively. For each point where both curves intersect, the polynomials f and
g have a common root and their resultant has to vanish. The resultant is a polynomial of degree mn in y. Each of its mn roots y corresponds to an x which can be determined by inversion using the same resultant.

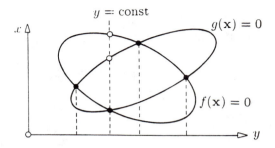

Figure 29.5: Curve-Curve intersection.

Surface-Surface intersection: Consider two algebraic surfaces in space given implicitly by $F(\mathbf{x}) = 0$ and $G(\mathbf{x}) = 0$, where

$$F = F_0 + F_1 x + \cdots + F_m x^m , \qquad G = G_0 + G_1 x + \cdots + G_n x^n,$$

and the F_i and G_i are polynomials in y and z. Their resultant R is a polynomial of degree mn in y and z. For each point where both surfaces intersect the resultant has to vanish. Hence, $R(y, z) = 0$ represents a cylinder in the direction of x containing the intersection curve. Similarly one can procure analogous cylinders in the y- and z-directions.

Another method for computing the intersection point by point is to use a surface $z = x(x, y)$. Inserting z into $F = 0$ and $G = 0$ gives the equations of two plane curves in the xy-plane. The intersection of these curves can be computed as described above. For varying $z(x, y)$ the intersection point of both surfaces will trace out the intersection curve. Often one chooses $z = z(x, y)$ to be a plane.

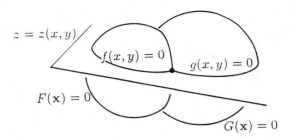

$z = z(x,y)$

$f(x,y) = 0$

$g(x,y) = 0$

$F(\mathbf{x}) = 0$

$G(\mathbf{x}) = 0$

Figure 29.6: Surface-Surface intersection, second method.

Curve-Surface and **Surface-Surface-Surface** intersections: The intersection of a curve, given as the intersection of two surfaces with a third surface, can be computed as the intersection of three surfaces. Let the three surfaces be given by $F = 0$, $G = 0$, $H = 0$. Writing

$$F = F_{00} + F_{10}x + \cdots + F_{mn}x^m y^n,$$

such that the F_{ij} only depend on z, and G and H in analogous form, one has that the resultant of F, G and H is a polynomial in z. Each root z of the resultant corresponds to a pair z, y which can be determined by inversion using the same resultant.

29.7 Notes and Problems

1 The monomial form of a polynomial curve in homogeneous coordinates can be written as

$$\mathbf{x}(t) = \mathbf{c}_0 \binom{n}{0} + \mathbf{c}_1 \binom{n}{1} t + \cdots + \mathbf{c}_n \binom{n}{n} t^n .$$

2 The monomial form of $\mathbf{x}(t)$ above can be converted to the Bézier representation by the rational parameter transformation

$$t = \frac{u}{1-u}$$

which maps $u = 0, \frac{1}{2}, 1$ to $t = 0, 1, \infty$. With this transformation one gets the Bézier representation

$$\varkappa(t) \cdot \varrho = \mathfrak{o}_0 \binom{n}{0} (1-u)^n + \mathfrak{o}_1 \binom{n}{1} (1-u)^{n-1} u + \cdots + \mathfrak{o}_n \binom{n}{n} u^n$$

$$= \mathfrak{o}_0 B_0^n(u) + \mathfrak{o}_1 B_1^n(u) + \cdots + \mathfrak{o}_n B_n^n(u) ,$$

where $\varrho = (1-u)^n$.

3 In homogeneous coordinates the rth step of the de Casteljau algorithm can be written as

$$\mathfrak{o}_j^r = \mathfrak{o}_j^{r-1} + \mathfrak{o}_{j+1}^{r-1} t .$$

4 A triangular quadratic surface represents a non-degenerate quadric if the conic sections corresponding to the lines in the parameter domain meet at one point of the surface.

5 In Note 4, it suffices to check the property for three lines of the parameter plane. Note that the curve tangents in the common point have to lie in one plane.

6 Consider a quadric Q given in homogeneous coordinates by its equation

$$Q_{\varkappa\varkappa} = \varkappa^t C \varkappa = 0 ,$$

a point $\mathfrak{c}$ on Q, and a further point $y \neq \varkappa$. In general, the ray from y to $\mathfrak{c}$ meets Q at a further point

$$\varkappa = \mathfrak{c}\lambda + y\mu ,$$

where

$$Q_{\varkappa\varkappa} = \lambda^2 Q_{\mathfrak{cc}} + 2\lambda\mu Q_{\mathfrak{cy}} + \mu^2 Q_{yy} = 0 .$$

Thus, one gets

$$\varkappa = \mathfrak{c}Q_{yy} - y2Q_{\mathfrak{cy}}$$

which is quadratic in y. In particular, let y run on a curve or surface of degree n, then the point $\varkappa$ traces out a curve or surface patch of degree $2n$ on Q.

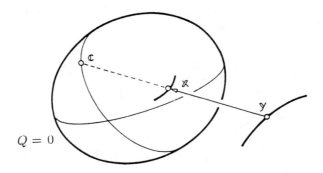

$Q = 0$

Figure 29.7: General stereographic projection

7 A (rational) cubic represents a conic section if it has one base point. This follows from Sections 6.8(6) and 25.7.

PART EIGHT

Differential Geometry

The origin of differential geometry lies mainly in the works of Leonid Euler (1707–1783), Gaspard Monge (1746–1818) and his disciple Dupin (1784–1873), and Carl Friedrich Gauss (1777–1855). Quite different from the approaches taken in the previous chapters, classical differential geometry studies curves and surfaces only with regard to their local properties. Methods of using differential calculus concepts are investigated, including tangency, curvature, and contact of some order. A crucial tool is the use of a local coordinate system. An infinitesimally small change of this local system along the curve or surface is expressed in the initial system.

The properties obtained by a local analysis also lead to results about the global nature of curves and surfaces. Examples of global structures are lines of curvatures, geodesic nets, and isometric maps which leave measurements on a surface invariant.

30 Curves

In this chapter intrinsic properties of smooth curves in 3-dimensional Euclidean space, such as arc length, curvature, and torsion, are discussed. Then local properties of curves, particularly the contact of order r of two curves, are studied. The main tools for such investigations are the use of a local coordinate system and the Frenet-Serret formulas.

Literature: do Carmo, Guggenheimer, Haack, Nutbourne·Martin

30.1 Parametric Curves and Arc Length

A parametric curve in $\mathcal{E}^3$ is given by

$$\mathbf{x} = \mathbf{x}(t) = \begin{bmatrix} x(t) \\ y(t) \\ z(t) \end{bmatrix} , \qquad t \in [a, b] \subset \mathbb{R} ,$$

where $x(t), y(t), z(t)$ are differentiable functions in t. For simplicity, only curves where

$$\dot{\mathbf{x}}(t) = \begin{bmatrix} \dot{x}(t) \\ \dot{y}(t) \\ \dot{z}(t) \end{bmatrix} \neq \mathbf{o} , \qquad t \in [a, b] ,$$

are considered. Such a parametrization is called **regular**. Any differentiable change of the parameter $\tau = \tau(t)$ does not change the curve. Moreover, if

$\dot{\tau} \neq 0$ in $[a, b]$, then $\mathbf{y}(t) = \mathbf{x}(\tau(t))$ also is a regular parametrization and $\tau(t)$ is invertible. A curve which allows for a regular parametrization is called **regular**.

Since $\mathbf{x}(t + \triangle t) = \mathbf{x}(t) + \dot{\mathbf{x}}(t)\triangle t + \cdots$, the vector $\dot{\mathbf{x}}$ spans the direction of the tangent of $\mathbf{x}$ at t. Moreover, the length of $\dot{\mathbf{x}}(t)\triangle t$ is an approximation of the length of the curve segment from t to $t + \triangle t$; $|\dot{\mathbf{x}}(t)|\, dt$ represents the **arc element** of the curve and

$$ s = s(t_0) = \int_a^{t_0} |\dot{\mathbf{x}}(t)|dt $$

is the **arc length** of the curve segment $\mathbf{x}[a, t_0]$. The arc length is independent of the parametrization and is itself the natural Euclidean parametrization of the curve. Namely on reparametrization by $t = t(s)$ and with $\mathbf{x} = \mathbf{x}(t(s))$ as a function of s one has

$$ \mathbf{x}'ds = \dot{\mathbf{x}}dt\ , $$

where the prime denotes differentiation with respect to the arc length s.

Example 1: Consider the helix

$$ x = \cos t\ , \qquad y = \sin t\ , \qquad z = t\ , \qquad t \in [0, b]\ . $$

Its tangent vector is

$$ \dot{x} = -\sin t\ , \qquad \dot{y} = \cos t\ , \qquad \dot{z} = 1\ , $$

and its arc length is $s = \sqrt{2}\, t$.

Remark 1: In most cases the arc length is not as simple as in Example 1 and has to be approximated by numerical methods.

30.2 The Frenet Frame

The main tool for discovering local properties of a curve is the **Frenet frame**, which is a local system closely related to the curve at some point

under consideration. Assuming sufficient differentiability, the first terms of
the Taylor expansion of $\mathbf{x}$ at t are

$$\mathbf{x}(t + \Delta t) = \mathbf{x} + \dot{\mathbf{x}}\Delta t + \ddot{\mathbf{x}}\frac{1}{2}\Delta t^2 + \dddot{\mathbf{x}}\frac{1}{6}\Delta t^3 + \cdots$$

where $\mathbf{x} = \mathbf{x}(t)$, etc. If the first three derivatives are linearly indepen-
dent, $\mathbf{x}; \dot{\mathbf{x}}, \ddot{\mathbf{x}}, \dddot{\mathbf{x}}$ forms a local affine system. In this system, the curve is
represented by its **canonical coordinates**

$$x = \Delta t + \cdots, \quad y = \frac{1}{2}\Delta t^2 + \cdots, \quad z = \frac{1}{6}\Delta t^3 + \cdots$$

where $+ \cdots$ stands for terms of degree higher than 3 in Δt.

From this local affine system one obtains a local Cartesian system by orthonormalization, namely

$$\mathbf{x}; \quad \mathbf{t} = \frac{\dot{\mathbf{x}}}{|\dot{\mathbf{x}}|}, \quad \mathbf{m} = \mathbf{b} \wedge \mathbf{t}, \quad \mathbf{b} = \frac{\dot{\mathbf{x}} \wedge \ddot{\mathbf{x}}}{|\dot{\mathbf{x}} \wedge \ddot{\mathbf{x}}|},$$

where "$\wedge$" denotes the alternating, i.e., cross product.

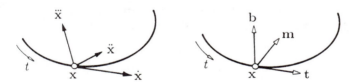

Figure 30.1: The local affine system and the Frenet frame.

The vector $\mathbf{t}$ is called the **tangent vector**, $\mathbf{m}$ is the **principal normal vector**,
and $\mathbf{b}$ is the **binormal vector**. The triple $\mathbf{t}, \mathbf{m}, \mathbf{b}$ is called the **Frenet frame**
of the curve at the point $\mathbf{x}$ under consideration. Note that the frame is
also defined if $\dddot{\mathbf{x}} = \mathbf{o}$.

The plane spanned by $\mathbf{x}; \dot{\mathbf{x}}, \ddot{\mathbf{x}}$ or $\mathbf{x}; \mathbf{t}, \mathbf{m}$ is called the **osculating plane**. Its
equation is

$$\mathbf{b}^t [\mathbf{x} - \mathbf{p}] = 0 ,$$

where $\mathbf{p} = \mathbf{x} + [\dot{\mathbf{x}} \ \ddot{\mathbf{x}}]\,\mathbf{y}$ represents any of its points. The plane spanned by $\mathbf{x}; \mathbf{t}, \mathbf{b}$ is called the **rectifying plane**, and the plane spanned by $\mathbf{x}; \mathbf{m}, \mathbf{b}$ is called the **normal plane** of the curve at $\mathbf{x}$.

30.3 Moving the Frame

Moving the Frenet frame along the curve reveals the behavior of the curve. Fundamental ideas of differential geometry are expressing the local change of this frame with respect to itself and using the arc length as parametrization. Some simple calculations procure the so-called **Frenet-Serret formula**

$$\begin{bmatrix} \mathbf{t}' & \mathbf{m}' & \mathbf{b}' \end{bmatrix} = \begin{bmatrix} \mathbf{t} & \mathbf{m} & \mathbf{b} \end{bmatrix} \begin{bmatrix} 0 & -\kappa & 0 \\ \kappa & 0 & -\tau \\ 0 & \tau & 0 \end{bmatrix} \, ,$$

where

$$\kappa = |\mathbf{x}''| = \frac{|\dot{\mathbf{x}} \wedge \ddot{\mathbf{x}}|}{|\dot{\mathbf{x}}|^3}$$

is called the **curvature** and

$$\tau = \frac{1}{\kappa^2} \det \begin{bmatrix} \mathbf{x}' & \mathbf{x}'' & \mathbf{x}''' \end{bmatrix} = \frac{\det \begin{bmatrix} \dot{\mathbf{x}} & \ddot{\mathbf{x}} & \dddot{\mathbf{x}} \end{bmatrix}}{|\dot{\mathbf{x}} \wedge \ddot{\mathbf{x}}|^2}$$

is called the **torsion** of the curve. Figure 30.2 illustrates this formula. For an interpretation using volumes see Section 16.7 and Note 6. Note that κ is zero at a point where $\ddot{\mathbf{x}}$ is parallel to $\dot{\mathbf{x}}$ and positive otherwise.

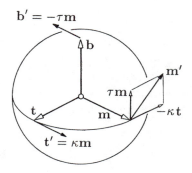

Figure 30.2: Frenet-Serret formula.

Remark 2: Torsion and the non-vanishing curvature of a curve as functions of the arc length are called its **natural equations**. They determine the curve uniquely up to its position in space.

Remark 3: Repeated differentiation of Frenet's formula gives

$$[\mathbf{x}' \quad \mathbf{x}'' \quad \mathbf{x}''' \quad] = [\mathbf{t} \quad \mathbf{m} \quad \mathbf{b}] \begin{bmatrix} 1 & 0 & -\kappa^2 & \cdots \\ 0 & \kappa & \kappa' & \cdots \\ 0 & 0 & \kappa\tau & \cdots \end{bmatrix}$$

from which one can read off the coordinate transformation from the canonical affine system to the Cartesian system $\mathbf{x}(s); \mathbf{t}, \mathbf{m}, \mathbf{b}$, see Figure 30.1. It follows that

$$\mathbf{x}(s + \triangle s) = \mathbf{x} + [\mathbf{t} \quad \mathbf{m} \quad \mathbf{b}] \begin{bmatrix} 1 & & -\frac{1}{6}\kappa^2 \\ \frac{1}{2}\kappa & & +\frac{1}{6}\kappa' \\ & & +\frac{1}{6}\tau\kappa \end{bmatrix} \begin{bmatrix} \triangle s \\ \triangle s^2 \\ \triangle s^3 \end{bmatrix} + \cdots$$

which is called the **canonical representation** of the curve at s.

30.4 The Spherical Image

Curvature and torsion have a concrete geometric meaning: The points $\mathbf{t}$ and $\mathbf{b}$ represent two curves on the unit sphere. Let α and γ denote their arc lengths respectively, then the Frenet-Serret formula gives

$$\kappa = \frac{d\alpha}{ds} \quad \text{and} \quad \tau = \pm\frac{d\gamma}{ds} .$$

Since α and γ represent the angles which $\mathbf{t}$ and $\mathbf{b}$ have turned through with s, one has: Curvature and torsion represent the **angular velocities** of the tangent and the osculating plane, respectively.

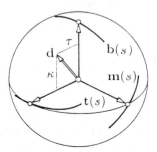

Figure 30.3: Spherical image and Darboux vector.

An infinitesimally small move of the Frenet frame with respect to s consists of a translation in the direction of $\mathbf{t}$ and a rotation. Since $\mathbf{t}', \mathbf{m}'$ and $\mathbf{b}'$ are orthogonal to $\mathbf{d} = \tau\mathbf{t} + \kappa\mathbf{b}$, the rotation axis is given by $\mathbf{x} + \lambda\mathbf{d}$. Moreover, $\mathbf{m}$ is orthogonal to $\mathbf{d}$ and of unit length. Therefore one gets for the angular velocity ω of the rotation

$$\omega = |\mathbf{m}'| = \sqrt{\kappa^2 + \tau^2}$$

and consequently $\omega = |\mathbf{d}|$. The vector $\mathbf{d}$ is called the **Darboux vector**.

Remark 4: Let

$$\mathbf{p} = \mathbf{x} + \mathbf{v} , \quad \mathbf{v} = [\mathbf{t} \quad \mathbf{m} \quad \mathbf{b}]\mathbf{y} ,$$

$\mathbf{y}$ fixed, denote a point moving with the frame. Since $|\mathbf{d}| = \omega$ one gets (see also Note 2)

$$\mathbf{p}' = \mathbf{t} + \mathbf{d} \wedge \mathbf{v} .$$

30.5 Osculating Plane and Sphere

A curve $\mathbf{x}(t)$ is said to have a **contact of order** r with a surface $F(\mathbf{x}) = 0$ at $t = t_0$ if

$$f(t_0) = 0 , \quad \dot{f}(t_0) = 0 , \quad , \quad f^{[r]}(t_0) = 0 ,$$

where $f(t) = F(\mathbf{x}(t))$. The definition of the order of contact does not depend on the parametrization $\mathbf{x}$. Namely, if $t = t(u)$ is a regular C^r-reparametrization, then $f(t) = \cdots = f^{[r]}(t) = 0$ at $t = t(u_0)$ holds if and only if $f(t(u_0)) = \cdots = f^{[r]}(t(u_0)) = 0$.

Geometrically, one has that $\mathbf{x}$ and F have contact of order r if the surface F contains $r + 1$ points infinitesimally close to the curve. More precisely, let $\mathbf{x}_i(t)$ be a sequence of curves with uniformly converging derivatives up to order r, where each $\mathbf{x}_i$ intersects F at $r + 1$ distinct points $\mathbf{x}_i(t_{ik}^0)$, $k = 0, 1, \ldots, r$, such that

$$\lim_{i \to \infty} t_{ik}^0 = t_0 \quad \text{for all } k$$

and

$$k.(t_0) \lim_{i \to \infty} \mathbf{x}_i^{[k]}(t_0) = \mathbf{x}^{[k]}(t_0) \ .$$

Then the functions $f_i(t) = F(\mathbf{x}_i(t))$ have $r + 1$ zeros which converge to t_0, and by Rolle's theorem $f_i^{[j]}(t)$ has $r + 1 - j$ zeros denoted by $t_{i0}^j, \ldots, t_{i,r-j}^j$ which also converge to t_0. Thus one has

$$f^{[j]}(t_0) = \lim_{i \to \infty} f(t_{i0}^j) = 0 \ , \qquad j = 0, \ldots, r \ ,$$

where $\mathbf{x} = \lim \mathbf{x}_i$.

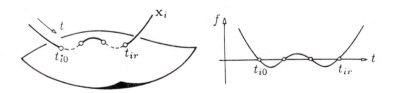

Figure 30.4: Contact of order r, definition.

From the Taylor expansion of $f(t)$ at a point of contact of order r one can deduce that the curve $\mathbf{x}(t)$ crosses the surface $F = 0$ at this point if r is even and that the curve stays on one side of the surface if r is odd (provided that r is the maximum order of contact, of course).

Example 2: Consider a plane given by $F = \mathbf{u}^t \mathbf{x} + u_0 = 0$. A curve $\mathbf{x}(t)$ contacts this plane with order 2 at t_0 if

$$f = 0 , \quad \dot{f} = \mathbf{u}^t \dot{\mathbf{x}} = 0 , \quad \ddot{f} = \mathbf{u}^t \ddot{\mathbf{x}} = 0 .$$

Obviously, the osculating plane spanned by $\mathbf{x}; \dot{\mathbf{x}}, \ddot{\mathbf{x}}$ satisfies these conditions.

Example 3: Consider a sphere given by $F = [\mathbf{x} - \mathbf{q}]^t [\mathbf{x} - \mathbf{q}] - R^2 = 0$. A curve $\mathbf{x}(t)$ contacts this sphere with order 3 at t if

$$f = 0 , \quad \dot{f} = 2[\mathbf{x} - \mathbf{q}]^t \dot{\mathbf{x}} = 0 , \quad \ddot{f} = 2(\dot{\mathbf{x}}^t \dot{\mathbf{x}} + [\mathbf{x} - \mathbf{q}]\ddot{\mathbf{x}}) = 0 ,$$

$$\dddot{f} = 2(3\dot{\mathbf{x}}^t \ddot{\mathbf{x}} + [\mathbf{x} - \mathbf{q}]^t \dddot{\mathbf{x}}) = 0 .$$

Let $\varrho = 1/\kappa$, then one can check that the sphere defined by

$$\mathbf{q} = \mathbf{x} + \mathbf{m}\varrho + \mathbf{b}\varrho'/\tau , \qquad R^2 = \varrho^2 + (\varrho')^2/\tau^2$$

satisfies these conditions with respect to arc length parametrization. It is called the **osculating sphere** of $\mathbf{x}$ at s.

Remark 5: The intersection of the osculating plane and the osculating sphere at some point $\mathbf{x}$ of a curve is called the **osculating circle** or **circle of curvature** of the curve at $\mathbf{x}$. Its center is given by

$$\mathbf{c} = \mathbf{x} + \mathbf{m}\varrho .$$

Its radius $\varrho = 1/\kappa$ is called the **radius of curvature** at $\mathbf{x}$.

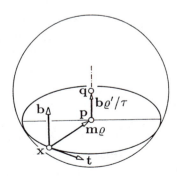

Figure 30.5: Osculating plane, circle, and sphere.

30.6 Osculating Curves

The notion of a contact of order r can be carried over to two curves. Let $\mathbf{x}(t)$ be one curve and let the other curve be given as the intersection of a number of surfaces. Then both curves have contact of order r at a common point if $\mathbf{x}(t)$ has at least the same order of contact at this point with each of the surfaces defining the other curve.

If both curves are given parametrically by $\mathbf{x}(t)$ and $\mathbf{y}(s)$, then these curves have contact of order r at a common point $\mathbf{x}(t_0) = \mathbf{y}(s_0)$ if there exists a regular reparametrization $t = t(u)$, where t_0 corresponds to u_0 such that the derivatives of both curves up to order r agree at the common point, i.e.,

$$\mathbf{y} = \mathbf{x}, \quad \frac{d\mathbf{y}}{ds} = \frac{d\mathbf{x}}{du}, \quad \ldots, \quad \frac{d^r\mathbf{y}}{ds^r} = \frac{d^r\mathbf{x}}{du^r}$$

at s_0 and u_0, respectively. These conditions can be rephrased by means of the chain rule. Let

$$\alpha = \frac{dt}{du}\Big|_{u_0}, \quad \beta = \frac{d^2t}{du^2}\Big|_{u_0}, \quad \gamma = \frac{d^3t}{du^3}\Big|_{u_0},$$

then

$$\mathbf{y} = \mathbf{x}, \quad \mathbf{y}' = \dot{\mathbf{x}}\alpha, \quad \mathbf{y}'' = \dot{\mathbf{x}}\beta + \ddot{\mathbf{x}}\alpha^2, \quad \mathbf{y}''' = \dot{\mathbf{x}}\gamma + \ddot{\mathbf{x}}3\alpha\beta + \dddot{\mathbf{x}}\alpha^3, \quad \ldots,$$

where the derivatives with respect to s and t are denoted by primes and , respectively. Thus, using matrix notation, a contact of order r at s_0 and t_0 means that

$$[\mathbf{y} \quad \mathbf{y}' \ldots \mathbf{y}^{(r)}] = [\mathbf{x} \quad \dot{\mathbf{x}} \ldots \mathbf{x}^{[r]}] \begin{bmatrix} 1 & & & \\ & \alpha & \cdots & * \\ & & \ddots & \vdots \\ & & & \alpha^r \end{bmatrix}$$

which is abbreviated by $Y = XC$. The entries $*$ of C are defined by the chain rule. The matrices Y and X are called the r-**jets** of $\mathbf{x}$ and $\mathbf{y}$, and C is referred to as the **chain rule connection matrix**, also cf. Note 8.

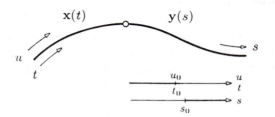

Figure 30.6: Osculating curves.

Example 4: The osculating circle which is defined in Remark 5 has contact of order 2 with the curve, since the curve has contact of order 2 with the osculating plane and contact of order 3 with the osculating sphere.

Example 5: In case of a contact of order 3 one has

$$[\mathbf{y} \quad \mathbf{y}' \quad \mathbf{y}'' \quad \mathbf{y}'''] = [\mathbf{x} \quad \dot{\mathbf{x}} \quad \ddot{\mathbf{x}} \quad \dddot{\mathbf{x}}] \begin{bmatrix} 1 & & & \\ & \alpha & \beta & \gamma \\ & & \alpha^2 & 3\alpha\beta \\ & & & \alpha^3 \end{bmatrix}.$$

Note that in this case, the curvatures, their first derivatives, and the torsions of both curves agree at $u = u_0$.

Remark 6: Note that contact of order r implies that curvature and torsion of both curves agree up to their $(r-2)$th and $(r-3)$th derivatives, respectively.

30.7 Notes and Problems

1 The direction of the principal normal $\mathbf{m}$ can be written as

$$\mathbf{m}\varrho = \ddot{\mathbf{x}} \cdot \dot{\mathbf{x}}^t\dot{\mathbf{x}} - \dot{\mathbf{x}} \cdot \ddot{\mathbf{x}}^t\dot{\mathbf{x}} .$$

2 Consider a rotation around an axis spanned by $\mathbf{o}; \mathbf{v}$. If $|\mathbf{v}|$ equals the angular velocity, then one has for any point $\mathbf{y}$,

$$\dot{\mathbf{y}} = \mathbf{v} \wedge \mathbf{y} .$$

Note that a change in the sign of $\mathbf{v}$ changes the sign of $\dot{\mathbf{y}}$.

3 In particular, let the axis direction be given by $\mathbf{v} = a\boldsymbol{\alpha} + b\boldsymbol{\beta} + c\boldsymbol{\gamma}$. Then one gets for the rotation of the Cartesian system $\mathbf{a}, \mathbf{b}, \mathbf{c}$ itself

$$[\dot{\mathbf{a}} \quad \dot{\mathbf{b}} \quad \dot{\mathbf{c}}] = [\mathbf{a} \quad \mathbf{b} \quad \mathbf{c}] \begin{bmatrix} 0 & -\gamma & +\beta \\ +\gamma & 0 & -\alpha \\ -\beta & +\alpha & 0 \end{bmatrix}.$$

4 Consider a parametric curve such that $\dot{\mathbf{x}}_+(t) = \alpha \dot{\mathbf{x}}_-(t)$ at $t = 0$, $\alpha > 0$. Then $\mathbf{x}(u)$ is C^1-continuous with respect to

$$u = \begin{cases} t & \text{if } t \leq 0 \\ \alpha t & \text{if } t \geq 0. \end{cases}$$

Note that this reparametrization does not change the degree of a piecewise polynomial curve.

5 Consider a parametric curve such that $\dot{\mathbf{x}}_+(t) = \alpha \dot{\mathbf{x}}_-$ and $\ddot{\mathbf{x}}_+ = \alpha^2 \ddot{\mathbf{x}}_- + \beta \dot{\mathbf{x}}_-$ at $t = 0$. Then $\mathbf{x}(u)$ is C^2-continuous with respect to

$$u = \begin{cases} t & \text{if } t \leq 0 \\ \alpha t + \frac{1}{2}\beta t^2 & \text{if } t \geq 0 \end{cases}$$

or where

$$u = \begin{cases} t & \text{if } t \leq 0 \\ \dfrac{\alpha^2 t}{\alpha - (\beta/2)t} & \text{if } t \geq 0. \end{cases}$$

Note that the second reparametrization does not change the degree of a piecewise rational curve (Degen '86).

6 Besides its arc length, a curve spanning $\mathbb{R}^d$ has $d - 1$ **geometric invariants** $\kappa_1, \ldots, \kappa_{d-1}$ defined by

$$\begin{aligned} \kappa_1 &= vol_2\,[\mathbf{x}' \quad \mathbf{x}''] & &= \kappa, \\ \kappa_1{}^2 \kappa_2 &= vol_3\,[\mathbf{x}' \quad \mathbf{x}'' \quad \mathbf{x}'''] & &= \tau\kappa^2, \\ \kappa_1{}^3 \kappa_2{}^2 \kappa_3 &= vol_4\,[\mathbf{x}' \quad \mathbf{x}'' \quad \mathbf{x}''' \quad \mathbf{x}''''], \end{aligned}$$

etc. The **Frenet frame** of such a curve is the positively oriented orthonormal basis obtained from $\mathbf{x}', \ldots, \mathbf{x}^{(d)}$ by an orthonormalization procedure.

7 Let $\mathbf{b}_1, \ldots, \mathbf{b}_d$ be the Frenet frame of a curve spanning $\mathbb{R}^d$, then the **general Frenet-Serret formula** reads

$$[\mathbf{b}_1' \ \ldots \ \mathbf{b}_d'] = [\mathbf{b}_1 \ \ldots \ \mathbf{b}_d] \begin{bmatrix} 0 & \kappa_1 & & & \\ -\kappa_1 & 0 & \kappa_2 & & \\ & -\kappa_2 & 0 & \ddots & \\ & & \ddots & \ddots & \kappa_{d-1} \\ & & & -\kappa_{d-1} & 0 \end{bmatrix}.$$

8 A regular curve $\mathbf{x}(t)$ in $\mathbb{R}^d$ with derivatives up to order r in $[a, t_0]$ and $[t_0, b]$ has a continuous Frenet frame and continuous curvatures $\kappa_1, \ldots, \kappa_{r-1}$ if and only if

$$[\mathbf{x}_- \quad \dot{\mathbf{x}}_- \ \ldots \ \mathbf{x}_-^{[r]}] = [\mathbf{x}_+ \quad \dot{\mathbf{x}}_+ \ \ldots \ \mathbf{x}_+^{[r]}] \begin{bmatrix} 1 & & & \\ & \alpha & \cdots & * \\ & & \ddots & \vdots \\ & & & \alpha^r \end{bmatrix},$$

where the elements marked by the asterisks in the connection matrix are completely arbitrary and $\mathbf{x}_-^{[i]}$ and $\mathbf{x}_+^{[i]}$ refer to the ith derivatives of $\mathbf{x}$ at t_0 in $[a, t_0]$ and $[t_0, b]$, respectively. Such curves are said to be **Frenet frame continuous of order** r (see Dyn·Micchelli) or **geometrically continuous of order** r, or more concisely G^r-**continuous**.

9 The curvatures $\kappa_1, \ldots, \kappa_r$ of a Frenet frame continuous curve of order r are only continuous, in general.

10 Note that chain rule continuity of order r implies continuity of the geometric invariants $\kappa_1, \ldots, \kappa_{r-1}$ but not vice versa, if $r > 2$.

11 From its definition it follows that a contact of order r is projectively invariant. Moreover, Frenet frame continuity is also projectively invariant, although the curvatures are Euclidean quantities. In general, however, the connection matrix describing the Frenet frame continuity changes under a projection.

12 For $r = 1$ and $r = 2$, chain rule continuity means tangent and curvature continuity, respectively.

13 If $\dddot{\mathbf{x}} \equiv \mathbf{o}$, one has $\tau \equiv 0$, and the curve is planar.

31 Curves on Surfaces

A surface can be thought of as being generated by sweeping a flexible curve through space. Then the family of curves as well as the trajectories of their points provides a parametric system. Any curve on such a surface has properties which depend on the surface. In particular, there are curve properties which depend merely on the intrinsic measurements of the surface. For instance, geodesics are completely defined in terms of the intrinsic surface geometry. Other curve properties depend on the curvature of the surface in space, e.g., lines of curvature have such properties. Again, the analysis rests crucially on the use of a local frame.

Literature: Blaschke·Leichtweiß, do Carmo, Klingenberg

31.1 Parametric Surfaces and Arc Element

Let $\mathbf{x}$ be a parametric surface in $\mathcal{E}^3$

$$\mathbf{x} = \mathbf{x}(u, v) = \begin{bmatrix} x(u, v) \\ y(u, v) \\ z(u, v) \end{bmatrix} , \qquad \mathbf{u} = (u, v) \in [\mathbf{a}, \mathbf{b}] \subset \mathbb{R}^2 ,$$

where the coordinates x, y, z are differentiable functions of u and v. For simplicity, the parametrization is assumed to be such that

$$\mathbf{x}_u \wedge \mathbf{x}_v \neq \mathbf{o} \qquad \text{for } \mathbf{u} \in [\mathbf{a}, \mathbf{b}] ,$$

i.e., both families of lines $\mathbf{x}(u,v)$ where u or v are kept fixed are regular and are nowhere tangential to each other. Such a parametrization is called **regular**. Any differentiable change of the parameter $\mathbf{r} = \mathbf{r}(\mathbf{u})$ does not change the surface. Moreover, if $\det [\mathbf{r}_u \quad \mathbf{r}_v] \neq 0$ for all $\mathbf{u} \in [\mathbf{a}, \mathbf{b}]$, then $\mathbf{y}(\mathbf{u}) = \mathbf{x}(\mathbf{r}(\mathbf{u}))$ is also a regular parametrization and $\mathbf{r}(\mathbf{u})$ is locally invertible. A surface which allows a regular parametrization is called **regular**.

Figure 31.1: Parametric surface.

Every regular curve $\mathbf{u} = \mathbf{u}(t)$ in the domain describes a regular curve on the surface $\mathbf{x}(\mathbf{u}(t))$. The **arc element** of this curve is given by

$$ds^2 = |\dot{\mathbf{x}}|^2 dt^2 = \left(\mathbf{x}_u^t \mathbf{x}_u \dot{u}^2 + 2\mathbf{x}_u^t \mathbf{x}_v \dot{u}\dot{v} + \mathbf{x}_v^t \mathbf{x}_v \dot{v}^2 \right) dt^2 \ .$$

For historic reasons, the arc element is written as

$$ds^2 = E du^2 + 2F du dv + G dv^2,$$

where $d\mathbf{u} = [du \quad dv]^t$ are considered free affine parameters. It is called the **first fundamental form** of the surface while E, F, G are called the **first fundamental quantities**:

$$E = E(u,v) = \mathbf{x}_u^t \mathbf{x}_u \ ,$$
$$F = F(u,v) = \mathbf{x}_u^t \mathbf{x}_v \ ,$$
$$G = G(u,v) = \mathbf{x}_v^t \mathbf{x}_v \ .$$

The first fundamental form defines the intrinsic measurement of the surface. It does not depend on the curve $\mathbf{u}(t)$.

Figure 31.2: Curve on a surface.

The arc length of the surface curve $\mathbf{x}(\mathbf{u}(t))$ is given by

$$s = \int_a^t |\dot{\mathbf{x}}|\, dt = \int_a^t \sqrt{E\dot{u}^2 + 2F\dot{u}\dot{v} + G\dot{v}^2}\, dt \; .$$

In particular, for $v = v_0 =$ fixed and u variable the curve $\mathbf{x}(u, v_0)$ is called a u-**line** which is an **isoline** of the surface. The arc element and arc length of a u-line are given by

$$ds = \sqrt{E}\, du \quad \text{and, respectively,} \quad s = \int \sqrt{E}\, du \; .$$

Remark 1: In matrix notation the arc element takes on the form

$$ds^2 = d\mathbf{u}^{t} C\, d\mathbf{u} \; , \qquad C = \begin{bmatrix} \mathbf{x}_u^{t} \\ \mathbf{x}_v^{t} \end{bmatrix} \begin{bmatrix} \mathbf{x}_u & \mathbf{x}_v \end{bmatrix} = \begin{bmatrix} E & F \\ F & G \end{bmatrix} \; .$$

Remark 2: The vectors $\mathbf{x}_u$ and $\mathbf{x}_v$ are tangent vectors of the u- and v-line, respectively. Hence if $F = 0$ at some $\mathbf{u}_0$, the isolines are orthogonal there. Moreover, if $F \equiv 0$, the net of isolines is orthogonal everywhere.

Remark 3: The **area element** associated with the element $du\,dv$ of the domain is given by

$$dA = |\mathbf{x}_u du \wedge \mathbf{x}_v dv| = |\mathbf{x}_u \wedge \mathbf{x}_v|\, du\,dv = \sqrt{\det C}\, du\,dv \; .$$

Figure 31.3: Normal and area element.

Remark 4: Let du_1, du_2 define two directions $[\mathbf{x}_u\ \mathbf{x}_v]^t du_1$, $[\mathbf{x}_u\ \mathbf{x}_v]^t du_2$ at a point on the surface $\mathbf{x}(\mathbf{u})$, then the angle γ formed by these directions satisfies

$$ds_1 ds_2 \cos\gamma = E\,du_1\,du_2 + F(du_1 dv_2 + du_2 dv_1) + G\,dv_1 dv_2 \ .$$

More concisely, this is written as $ds_1 ds_2 \cos\gamma = d\mathbf{u}_1^t C\,d\mathbf{u}_2$.

31.2 The Local Frame

The partial derivatives $\mathbf{x}_u, \mathbf{x}_v$ span the direction of the tangent plane at $\mathbf{x}$. Let $\mathbf{y}$ denote a point of this plane, then

$$\det\,[\mathbf{y} - \mathbf{x}, \mathbf{x}_u, \mathbf{x}_v] = 0$$

is the equation of this plane. The normal of the surface at $\mathbf{x}$ is the normal $\mathbf{x}_u \wedge \mathbf{x}_v$ of the tangent plane. The normalized normal

$$\mathbf{n} = [\mathbf{x}_u \wedge \mathbf{x}_v]\,\frac{1}{\sqrt{\det C}}$$

together with the non-normalized vectors $\mathbf{x}_u$ and $\mathbf{x}_v$ form a **natural local frame** of the surface at $\mathbf{x}$. This frame plays the same important role that the Frenet frame does for curves. Note that this frame forms only an affine system which also depends on the parametrization of the surface.

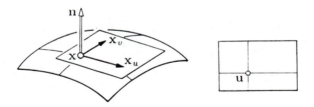

Figure 31.4: The local frame.

31.3 The Curvature of a Curve

Let $\mathbf{u}(t)$ define a curve on the surface $\mathbf{x}(\mathbf{u})$ and let the prime denote differentiation with respect to the arc length of $\mathbf{x}(\mathbf{u}(t))$. Then one has

$$\mathbf{t} = \mathbf{x}' = \mathbf{x}_u u' + \mathbf{x}_v v' \,,$$
$$\mathbf{t}' = \mathbf{x}'' = \mathbf{x}_{uu} u'^2 + 2\mathbf{x}_{uv} u' v' + \mathbf{x}_{vv} v'^2 + \mathbf{x}_u u'' + \mathbf{x}_v v'' \,.$$

Let φ be the angle between the principal normal $\mathbf{m}$ of the curve and the surface normal $\mathbf{n}$ at some $\mathbf{x}$. Then using the Frenet-Serret formula one gets

$$\mathbf{n}^t \mathbf{t}' = \mathbf{n}^t \mathbf{m} \kappa = \kappa \cos \varphi \,.$$

Inserting $\mathbf{t}'$ from above and using the fact that $\mathbf{n}$ is orthogonal to $\mathbf{x}_u$ and $\mathbf{x}_v$ gives

$$\mathbf{n}^t \mathbf{t}' = \mathbf{n}^t \mathbf{x}_{uu} u'^2 + 2\mathbf{n}^t \mathbf{x}_{uv} u' v' + \mathbf{n}^t \mathbf{x}_{vv} v'^2 \,.$$

Hence, one has

$$\kappa \cos \varphi = \mathbf{n}^t \mathbf{x}_{uu} u'^2 + 2\mathbf{n}^t \mathbf{x}_{uv} u' v' + \mathbf{n}^t \mathbf{x}_{vv} v'^2 \,.$$

This is commonly written as

$$\kappa \cos \varphi \, ds^2 = L \, du^2 + 2M \, du \, dv + N \, dv^2$$

and called the **second fundamental form** in classical differential geometry while, L, M, N are the so-called **second fundamental quantities**. Since $\mathbf{n}^t\mathbf{x}_u = 0$ implies $\mathbf{n}_u^t\mathbf{x}_u = -\mathbf{n}^t\mathbf{x}_{uu}$, etc., one has

$$
\begin{aligned}
L = L(u, v) &= \mathbf{n}^t\mathbf{x}_{uu} = -\mathbf{n}_u^t\mathbf{x}_u \\
M = M(u, v) &= \mathbf{n}^t\mathbf{x}_{uv} = -\mathbf{n}_v^t\mathbf{x}_u = -\mathbf{n}_u^t\mathbf{x}_v \\
N = N(u, v) &= \mathbf{n}^t\mathbf{x}_{vv} = -\mathbf{n}_v^t\mathbf{x}_v \ .
\end{aligned}
$$

The second fundamental form has the following geometric meaning: Given a point $\mathbf{x}$ on the surface, a tangent direction given by the ratio $du : dv$, and an angle φ, one can, with the aid of both fundamental forms, compute the curvature of a surface curve going through $\mathbf{x}$ and having the prescribed tangent and osculating plane.

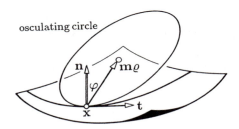

osculating circle

Figure 31.5: The second fundamental form, definition.

Remark 5: In matrix notation the second fundamental form takes on the form

$$
\kappa \cos \varphi \, ds^2 = d\mathbf{u}^t D d\mathbf{u} \ ,
$$

where

$$
D = -[\mathbf{n}_u \quad \mathbf{n}_v]^t[\mathbf{x}_u \quad \mathbf{x}_v] = \begin{bmatrix} L & M \\ M & N \end{bmatrix} \ .
$$

31.4 Meusnier's Theorem

The right-hand of the second fundamental form divided by ds^2 does not depend on φ. Hence it agrees with the curvature κ if $\varphi = 0°$. In this case the osculating plane of the curve contains the surface normal, which means that it is perpendicular to the surface at $\mathbf{x}$. This curvature is called the **normal curvature** κ_n of the surface at $\mathbf{x}$ in the direction of $\mathbf{t}$ and is given by

$$\kappa_n = \frac{1}{\varrho_n} = \frac{L\,du^2 + 2M\,du\,dv + N\,dv^2}{E\,du^2 + 2F\,du\,dv + G\,dv^2} = \frac{d\mathbf{u}^t D\,d\mathbf{v}}{d\mathbf{u}^t C\,d\mathbf{u}} \ .$$

Comparing this expression with the second fundamental form above gives

$$\kappa_n = \kappa\cos\varphi \quad \text{or} \quad \varrho = \varrho_n\cos\varphi \ ,$$

where $\varrho = 1/\kappa$. This simple formula has an interesting and important geometric interpretation, known as **Meusnier's theorem**:

> The osculating circles of all surface curves which have the same tangent at the common point $\mathbf{x}$ form a sphere. This sphere has a common tangent plane with the surface at $\mathbf{x}$, and it has radius ϱ_n.

This sphere is illustrated in Figure 31.6. Note that $0 \le \varrho \le |\varrho_n|$ and that $\varrho = 0$ if and only if $\varphi = 90°$, i.e., if the osculating plane coincides with the tangent plane.

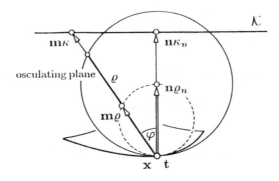

Figure 31.6: Meusnier's sphere.

Remark 6: It also follows that for fixed $\mathbf{t}$ and varying φ the points $\mathbf{k} = \mathbf{x} + \kappa\mathbf{m}$ lie on a straight line orthogonal to $\mathbf{t}$ and $\mathbf{n}$ with distance

κ_n from $\mathbf{x}$. This is of particular interest if the surface is a cylinder with generatrices orthogonal to $\mathbf{t}$.

31.5 The Darboux Frame

In some applications it is convenient to use the following local frame related to a surface and a curve on it. This frame is given by

$$\mathbf{t}\,, \quad \mathbf{n}\,, \quad \mathbf{s} = \mathbf{t}\wedge\mathbf{n} = \mathbf{m}\sin\varphi + \mathbf{b}\cos\varphi$$

and is called **Darboux frame**. It can be obtained from the Frenet frame of $\mathbf{x}(\mathbf{u}(t))$ by a rotation around $\mathbf{t}$.

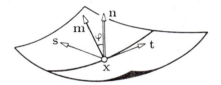

Figure 31.7: The Darboux frame.

Differentiating the frame and applying the Frenet-Serret formulas gives

$$[\mathbf{t}' \quad \mathbf{n}' \quad \mathbf{s}'] = [\mathbf{t} \quad \mathbf{n} \quad \mathbf{s}]\begin{bmatrix} 0 & -\kappa_n & -\kappa_g \\ \kappa_n & 0 & -\tau_g \\ \kappa_g & \tau_g & 0 \end{bmatrix},$$

where

$$\kappa_n = \kappa\cos\varphi\,, \quad \kappa_g = \kappa\sin\varphi\,, \quad \tau_g = \tau + \varphi'\,.$$

The quantity κ_n is called the **normal curvature** in the direction of $\mathbf{t}$, κ_g is called the **geodesic curvature**, and τ_g is called the **geodesic torsion** of the surface curve $\mathbf{x}(\mathbf{u}(t))$. They have the following geometric meaning:

Since $\mathbf{x}'' = \mathbf{t}' = \mathbf{s}\kappa_g + \mathbf{n}\kappa_n$, the normal curvature is also obtained by projecting the second derivative $\mathbf{x}''$ into the **normal section plane** spanned by $\mathbf{x}; \mathbf{t}, \mathbf{n}$. This means that, $\kappa_n \mathbf{n}$ is the projection of $\kappa \mathbf{m}$. Moreover, κ_n equals the curvature of the projection of the curve into the normal plane corresponding to $\mathbf{t}$.

If $\kappa_n = 0$, $\mathbf{t}$ is called **asymptotic**, and if, along the curve, $\kappa_n \equiv 0$, the curve is called an **asymptotic line**; its osculating planes are tangent planes of the surface.

Since $\mathbf{x}'' = \mathbf{t}' = \mathbf{s}\kappa_g + \mathbf{n}\kappa_n$, the geodesic curvature is obtained by projecting the second derivative $\mathbf{x}''$ into the tangent plane. This means that $\kappa_g \mathbf{s}$ is the projection of $\kappa \mathbf{m}$. Moreover, κ_g equals the curvature of the projection of the curve into the tangent plane. Note that $\kappa_g^2 + \kappa_n^2 = \kappa^2$.

If, along the curve, $\kappa_g \equiv 0$, the curve is called a **geodesic**. Since $\kappa_g = \kappa_g \det [\mathbf{t} \quad \mathbf{n} \quad \mathbf{s}] = -\det [\mathbf{x}' \quad \mathbf{x}'' \quad \mathbf{n}]$, the second derivative of a geodesic, the so-called **accelerative force**, is perpendicular to the surface. Thus, an object running on a surface traces out a geodesic if there is no external force.

Since $\mathbf{n}' = -\mathbf{t}\kappa_n + \mathbf{s}\tau_g$ and $\mathbf{s} = \mathbf{t} \wedge \mathbf{n}$, one has $\tau_g = \det [\mathbf{t} \ \mathbf{n} \ \mathbf{n}']$ which solely depends on $\mathbf{t}$. Along a geodesic line one has $\varphi \equiv 0$ and therefore $\tau_g = \tau$. This accounts for the name geodesic torsion.

If $\tau_g = 0$, $\mathbf{t}$ is called a **principal direction** and if $\tau_g \equiv 0$, the curve is called a **line of curvature**, see Section 32.1. Along a line of curvature, the tangent vector and the derivative of the surface normal are linearly dependent. Hence, the surface normals at $\mathbf{x}$ and at $\mathbf{x} + d\mathbf{x}$ are coplanar. Conversely, if this property holds for all points of the curve, the curve is a line of curvature.

31.6 Notes and Problems

1 Note that $ds^2 > 0$ holds for any real du, dv. However, $ds^2 = 0$ holds for two imaginary directions $dv/du = (-F \pm \sqrt{F^2 - EG})/G$, the so-called **isotropic** or **minimal directions**, cf. Section 23.7.

2 The derivatives of the local frame in Section 31.2 can be expressed by the frame itself. This establishes the **fundamental equations** of **Gauss**,

$$
\begin{bmatrix} \mathbf{x}_{uu} & \mathbf{x}_{uv} & \mathbf{x}_{vv} \end{bmatrix} = \begin{bmatrix} \mathbf{x}_u & \mathbf{x}_v & \mathbf{n} \end{bmatrix} \begin{bmatrix} a_{1,1} & a_{1,2} & a_{2,2} \\ b_{1,1} & b_{1,2} & b_{2,2} \\ L & M & N \end{bmatrix} ,
$$

and **Weingarten**,

$$
\begin{bmatrix} \mathbf{n}_u & \mathbf{n}_v \end{bmatrix} = \begin{bmatrix} \mathbf{x}_u & \mathbf{x}_v \end{bmatrix} \begin{bmatrix} p_1 & p_2 \\ q_1 & q_2 \end{bmatrix} .
$$

3 The coefficients $a_{1,1}, \ldots, b_{2,2}$ in Note 2 are called **Christoffel symbols** of the second kind. They only depend on E, F, G and their first partials.

4 The Weingarten matrix solves the linear system

$$
C \begin{bmatrix} p_1 & p_2 \\ q_1 & q_2 \end{bmatrix} = D .
$$

5 The direction $\mathbf{c} = \tau_g \mathbf{t} + \kappa_n \mathbf{s}$ is called **conjugate** to $\mathbf{t}$ at $\mathbf{x}$. One has $\mathbf{c}^t \mathbf{n}' = 0$.

6 The direction $\mathbf{t}$ is **self-conjugate** if $\kappa_n = 0$.

7 The pair $\mathbf{x}(t)$ and $\mathbf{n}(t)$ defines a so-called **strip**. In particular, it is called an **asymptotic** or **osculating strip** if $\kappa_n \equiv 0$, a **geodesic strip** if $\kappa_g \equiv 0$, and a **curvature strip**, if $\tau_g \equiv 0$.

8 There is only one asymptotic and only one geodesic strip through a given twisted curve. However, there are ∞^1 curvature strips through a given twisted curve, any two of them having a constant angle between their osculating planes.

9 The geodesic lines $\mathbf{x}(\mathbf{u}(s))$ form the solution of the linear differential equation $\kappa_g = -\det\begin{bmatrix} \mathbf{x}' & \mathbf{x}'' & \mathbf{n} \end{bmatrix} = 0$ which is of second order in u and v.

10 The curvature lines $\mathbf{x}(\mathbf{u}(s))$ form the solution of the quadratic differential equation $\kappa_g = \det\begin{bmatrix} \mathbf{x}' & \mathbf{n} & \mathbf{n}' \end{bmatrix}$ which is of first order in u and v.

32 Surfaces

One of the interesting properties of a curved surface is its curvature which is strongly related to a conic, the so-called Dupin indicatrix. Together with Meusnier's sphere this indicatrix completely describes the curvature properties at a point of the surface. Again, the use of a local frame is the crucial tool for analysis.

Literature: Blaschke·Leichtweiß, do Carmo, Klingenberg

32.1 Dupin's Indicatrix and Euler's Theorem

Let $\triangle \mathbf{u} = [\triangle u \quad \triangle v]^{\mathrm{t}}$ denote the coordinates of a point in the tangent plane at $\mathbf{x}$ with respect to the affine system $\mathbf{x}; \mathbf{x}_u, \mathbf{x}_v$, assume that L, M, N are not all zero, and let the sign of the normal $\mathbf{n}$ be such that

$$L\triangle u^2 + 2M\triangle u \triangle v + N\triangle v^2 = 1$$

represents a conic section with real points. This conic is called **Dupin's indicatrix**; it provides deeper insight into the local shape of the surface.

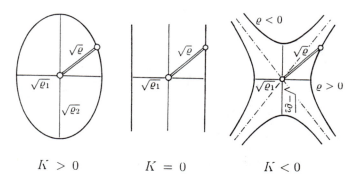

$$K > 0 \qquad\qquad K = 0 \qquad\qquad K < 0$$

Figure 32.1: Types of Dupin indicatrices.

Recall both fundamental forms: The distance $\triangle s$ between $\mathbf{x}$ and some point $\triangle \mathbf{u}$ satisfies $\triangle s^2 = \triangle \mathbf{u}^t C \triangle \mathbf{u}$. Hence for $\triangle \mathbf{u}$ on Dupin's indicatrix, one has

$$\kappa_n = \frac{1}{\varrho_n} = \frac{1}{\triangle s^2}$$

and therefore

$$\triangle s = \sqrt{\varrho_n} \ .$$

Consequently, the extrema ϱ_1 and ϱ_2 of ϱ_n as a function of $\triangle \mathbf{u}$ agree with the squared semi-axes of Dupin's indicatrix. In particular, for $\kappa_1 \kappa_2 > 0$ the indicatrix is an ellipse, and for $\kappa_1 \kappa_2 < 0$ it is one of two homothetic hyperbolas, depending on the sign of $\mathbf{n}$. For $\kappa_1 \kappa_2 = 0$ the indicatrix degenerates to a pair of lines.

Let ϑ denote the angle between the principal axis corresponding to κ_1 and the line joining $\mathbf{x}$ and $\triangle \mathbf{u}$. Taking the principal axes as basis directions such that $\triangle \mathbf{u}$ is represented by $[\triangle s \cos \vartheta \quad \triangle s \sin \vartheta]^t$, Dupin's indicatrix takes on the form

$$\frac{\cos^2 \vartheta}{\varrho_1} + \frac{\sin^2 \vartheta}{\varrho_2} = \frac{1}{\varrho} \ .$$

This formula known as **Euler's theorem** is illustrated in Figure 32.2 for an ellipse.

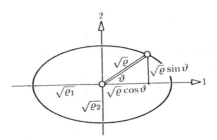

Figure 32.2: Euler's theorem, example.

The axes of Dupin's indicatrix can be found as follows. Consider the pencil spanned by the unit circle $\triangle \mathbf{u}^t C \triangle \mathbf{u} = 1$ and Dupin's indicatrix $\triangle \mathbf{u}^t D \triangle \mathbf{u} = 1$,

$$\triangle \mathbf{u}^t [D - \lambda C] \triangle \mathbf{u} = 1 - \lambda .$$

This pencil contains three pairs of lines as illustrated in Figure 32.3. A (non-interesting) line for $\lambda = 1$ and two pairs of parallel lines for the roots λ_1 and λ_2 of

$$\det [D - \lambda C] = 0 .$$

The corresponding axis directions are solutions of

$$[D - \lambda_i C] \triangle \mathbf{u}_i = \mathbf{o} , \quad i = 1, 2$$

and represent the **directions of principal curvature**. Moreover,

$$\triangle \mathbf{u}_i^t D \triangle \mathbf{u}_i = \lambda_i \cdot \triangle \mathbf{u}_i^t C \triangle \mathbf{u}_i$$

implies $\lambda_i = \kappa_i$. The curves whose tangent directions are directions of principal curvature are curvature lines as one can observe from the definition in Section 31.5 and Remark 5.

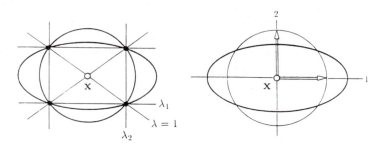

Figure 32.3: Unit circle and Dupin's indicatrix.

Remark 1: If $\kappa_1 = \kappa_2$, Dupin's indicatrix is a circle and the normal curvature is the same for all directions. Such a point is called an **umbilical point**; the principal directions are undefined there. At an umbilical point $\triangle \mathbf{u}^t D \triangle \mathbf{u}$ equals $\triangle \mathbf{u}^t C \triangle \mathbf{u}$ up to a factor, i.e., a point is an umbilical point if $E : F : G = L : M : N$.

Example 1: The curvature lines of a surface of revolution are the **meridians** and the **parallels**. Note that all points of a sphere are umbilical points.

32.2 Gaussian Curvature and Mean Curvature

As shown above, the principal curvatures κ_1 and κ_2 are the zeros of

$$\det [D - \kappa C] .$$

Since this expression is quadratic in κ, one has

$$\kappa^2 - (\kappa_1 + \kappa_2)\kappa + \kappa_1 \kappa_2 = \det [D - \kappa C] ,$$

where

$$\kappa_1 \kappa_2 = \frac{LN - M^2}{EG - F^2} , \qquad \kappa_1 + \kappa_2 = 2\,\frac{NE - 2MF + LG}{EG - F^2} .$$

The quantity $K = \kappa_1 \kappa_2$ is called the **Gaussian** or **total curvature**, and $H = \frac{1}{2}(\kappa_1 + \kappa_2)$ is called the **mean curvature**.

If $K > 0$, $K = 0$ or $K < 0$ at $\mathbf{x}$, the point $\mathbf{x}$ is called **elliptic**, **parabolic**, or **hyperbolic**, respectively, in accordance with the shape of Dupin's indicatrix. Figure 32.4 illustrates the three different kinds of points on a torus. Finally, if $\kappa_1 = \kappa_2 = 0$, as is the case for the points on a plane, then $K = H = 0$ and the point $\mathbf{x}$ is called a **flat point**.

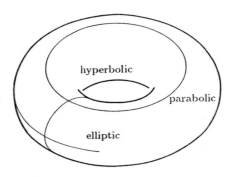

Figure 32.4: Elliptic, hyperbolic, and parabolic points on a torus.

Remark 2: The coefficients E, F, G of the first fundamental form do not change if the surface is deformed such that distances on the surface remain the same. Properties of a surface depending only on the first fundamental form but not on the embedding of the surface in space make up the so-called **intrinsic geometry** of a surface. One of the most important facts of differential geometry is that the total curvature K is also an intrinsic property. This fact, known as **theorema egregium**, is due to Gauss who proved in 1826 that K can be expressed in terms of E, F, G and their partial derivatives, see Note 6.

Remark 3: A surface is called a **developable** if it can be unrolled onto a plane without changing its intrinsic geometry. As a consequence of the theorema egregium the total curvature of a developable surface must be identically zero. Conversely, one can show that each surface with $K \equiv 0$ is a developable, see also Section 32.4.

Remark 4: The Gaussian curvature is related to the area of Dupin's indicatrix. If $K \geq 0$, the area equals $\pi/\sqrt{K}$. If $K < 0$, the area bounded by a tangent and both asymptotes equals $1/\sqrt{-K}$.

32.3 Conjugate Directions and Asymptotic Lines

Recall from Section 14.2 on affine quadrics that two directions $\triangle \mathbf{u}$ and $\triangle \mathbf{v}$ are **conjugate** with respect to Dupin's indicatrix if

$$\triangle \mathbf{u}^t D \triangle \mathbf{v} = 0 \ .$$

This relationship reflects an interesting geometric property:

If the point $\mathbf{x}$ moves by an infinitesimally small amount in the direction of $\triangle \mathbf{u}$, the tangent plane at $\mathbf{x}$ turns around a line in the direction of $\triangle \mathbf{v}$ conjugate to $\triangle \mathbf{u}$, as illustrated in Figure 32.5. Analytically one can prove this fact as follows. The common direction $\triangle \mathbf{y} = [\mathbf{x}_u \ \mathbf{x}_v] \triangle \mathbf{v}$ of two neighbored tangent planes satisfies $\triangle \mathbf{y}^t \mathbf{n} = 0$ and $\triangle \mathbf{y}^t \mathbf{n}' = 0$. Thus one has

$$\triangle \mathbf{y}^t \cdot \mathbf{n}' = \triangle \mathbf{v}^t [\mathbf{x}_u \ \ \mathbf{x}_v]^t \cdot [\mathbf{n}_u \ \ \mathbf{n}_v] \triangle \mathbf{u} = -\triangle \mathbf{v}^t D \triangle \mathbf{u} = 0 \ .$$

This agrees with the definition given in Notes 5 and 6 of Section 31.6 where $\triangle \mathbf{y} = \mathbf{c}$. In particular, $\triangle \mathbf{u}$ is **self-conjugate** or **asymptotic** if

$$\triangle \mathbf{u}^t D \triangle \mathbf{u} = 0 \ .$$

In general, there are two asymptotic directions at a point $\mathbf{x}$. They are real and distinct if $K < 0$, they coalesce if $K = 0$ and $H \neq 0$, and they are non-real if $K > 0$. They are not defined if $K = H = 0$.

Note that the normal curvature in an asymptotic direction is zero. Therefore, a curve on a surface enveloped by asymptotic directions is an asymptotic line. A u-line is an asymptotic line if $L \equiv 0$ along this line. Furthermore, if $L \equiv 0$ and $N \equiv 0$ for the entire surface, then the net of isolines forms a net of **asymptotic lines** on the surface. The net of curvature lines bisects the angles of the asymptotic net.

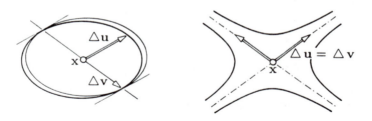

Figure 32.5: Conjugate and asymptotic directions.

Example 2: The asymptotic net on a quadric is formed by its (real or non-real) generatrices.

Remark 5: Since principal directions of curvature are conjugate and orthogonal, Section 31.6(5) implies that $\tau_g = 0$ along a curvature line.

Remark 6: The u- and v-lines are conjugate if $M = 0$ and they form a **conjugate net** if $M \equiv 0$.

Remark 7: The isolines form the net of curvature lines if they are orthogonal and conjugate, i.e., if $F \equiv 0$ and $M \equiv 0$.

Remark 8: The geometric properties of conjugate and asymptotic directions imply that these directions are **projectively invariant**.

Remark 9: Recall that the conjugate directions of a quadric are in harmonic position to the asymptotic directions at the same point and that the directions of principal curvature are in harmonic position both to the asymptotic and to the isotropic directions.

32.4 Ruled Surfaces and Developables

A straight line $\mathbf{x} = \mathbf{p} + \mathbf{v}r$ being moved through space sweeps out a **ruled surface**,

$$\mathbf{x}(r, t) = \mathbf{p}(t) + \mathbf{v}(t) \cdot r \ .$$

The curve $\mathbf{p}(t)$ is called a **directrix**. The r-lines are called **generatrices**. Since the partials are

$$\mathbf{x}_t = \dot{\mathbf{p}} + \dot{\mathbf{v}}r \qquad \text{and} \qquad \mathbf{x}_r = \mathbf{v} \ ,$$

the normal at $\mathbf{x}$ is given by

$$\mathbf{n}\varrho = \mathbf{a} + \mathbf{b}r \ ,$$

where $\mathbf{a} = [\dot{\mathbf{p}} \wedge \mathbf{v}]$, $\mathbf{b} = [\dot{\mathbf{v}} \wedge \mathbf{v}]$, and ϱ is such that $|\mathbf{n}| = 1$. Thus, a point $\mathbf{y}$ of the tangent plane at $\mathbf{x}$ satisfies

$$\mathbf{a}^t [\mathbf{y} - \mathbf{p}] + r\mathbf{b}^t [\mathbf{y} - \mathbf{p}] = 0 \ .$$

This means geometrically that the tangent planes along a generatrix form a pencil with r as a natural parameter.

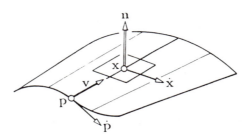

Figure 32.6: Ruled surface.

However, if for some t, $\mathbf{a}$ and $\mathbf{b}$ are linearly dependent, i.e., if

$$\det [\dot{\mathbf{p}} \quad \mathbf{v} \quad \dot{\mathbf{v}}] = 0 \ ,$$

then there is only one tangent plane along the corresponding generatrix. Consequently, if $\mathbf{a}$ and $\mathbf{b}$ are linearly dependent for all t, each tangent plane depends only on t. In this case, the tangent planes of the entire surface form a one-parameter family of planes.

Such a surface with a one-parameter family of tangent planes is a developable, see Remark 3. Conversely, any one-parameter family of planes

$$\mathbf{u}^t(t)\mathbf{y} + u_0(t) = 0$$

envelopes a developable. Moreover, a general developable surface is enveloped by such a family of planes where two consecutive planes meet in a generatrix and three consecutive planes meet in a (possibly ideal) point. Hence, a developable surface is the union of cones ($\mathbf{p} = $ *fixed*), cylinders ($\mathbf{v} = $ *fixed*), and surfaces formed by the tangents of a space curve ($\mathbf{v} = \dot{\mathbf{p}}$).

Both families of asymptotic lines of a developable coalesce and coincide with the family of generatrices and also with one family of curvature lines. As a consequence, the second family of curvature lines is formed by the trajectories orthogonal to the generatrices. Note that these orthogonal trajectories are equidistant.

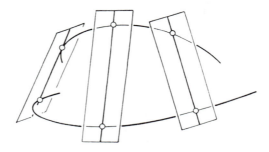

Figure 32.7: Developable.

Example 3: The tangent surface of a curve $\mathbf{p}(t)$ is given by

$$\mathbf{x}(r,t) = \mathbf{p}(t) + \dot{\mathbf{p}}(t)(r - t) .$$

Example 4: The common tangent planes of two space curves also envelope a developable. This is illustrated in Figure 32.7.

Remark 10: The condition $\tau_g = -\det[\mathbf{x}'\ \ \mathbf{n}\ \ \mathbf{n}'] = 0$ along a surface curve is just the condition for the normals of a curvature line to form a developable.

32.5 Contact of Order r

The notion of contact of order r introduced in Sections 30.5 and 30.6 can be extended to surfaces. Two sufficiently often differentiable surfaces have contact of order r at a common point $\mathbf{x}_0$ if their intersections with an arbitrary plane through $\mathbf{x}_0$ have contact of order r. This definition can be generalized to a contact of two surfaces along an entire curve $\mathbf{p}(t)$, as illustrated in Figure 32.8.

In particular, two abutting surfaces $\mathbf{x}$ and $\mathbf{y}$ have **contact of order** 0 if their common boundary curve $\mathbf{p}(t)$ is continuous.

The surfaces $\mathbf{x}$ and $\mathbf{y}$ have **contact of order** 1 along a common curve $\mathbf{p}(t)$ if the tangent planes at every $\mathbf{p}(t)$ agree. If $\mathbf{p}$ is a C^1-curve, both normal curvatures in the direction of $\mathbf{t} = \dot{\mathbf{p}}$ and both directions conjugate to $\mathbf{t}$ agree for all t. This implies that the two indicatrices of $\mathbf{x}$ and $\mathbf{y}$ at every $\mathbf{p}(t)$ have common points in the direction of $\pm\mathbf{t}$ and also common tangents there.

In particular, the pairs of indicatrices agree, if there is a family of C^2 curves across the common curve $\mathbf{p}(t)$.

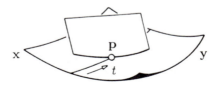

Figure 32.8: Contact of order r, definition.

The surfaces $\mathbf{x}_-$ and $\mathbf{x}_+$ have **contact of order** 2 along a common curve $\mathbf{p}(t)$ if their indicatrices at $\mathbf{p}(t)$ agree for all t. Hence, if $\mathbf{p}$ is a piecewise C^1-curve it suffices to check that the pairs of indicatrices agree in one further point.

Remark 11: Recall that contact of order r for curves means C^r-continuity with respect to some suitable parametrization. The same is true for surfaces (Herron '87).

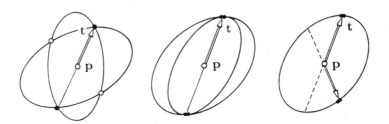

Figure 32.9: Contact of order 1 and 2 and Dupin's indicatrices.

Remark 12: In the special case where the test plane has contact of order s with the curve of contact itself, both plane intersections have contact of order $(r+1)(s+1) - 1$.

Example 5: If two surfaces have first order contact along a C^1-curve $\mathbf{p}(t)$, where $\dot{\mathbf{p}}$ is an asymptotic direction, both Dupin indicatrices have an ideal intersection of multiplicity 4 in the direction of $\dot{\mathbf{p}}$.

32.6 Notes and Problems

1 The paraboloid defined by $2\triangle\tau = L\triangle u^2 + 2M\triangle u\triangle v + N\triangle v^2$ in the local system $\mathbf{x}; \mathbf{x}_u, \mathbf{x}_v, \mathbf{n}$ is called the **osculating paraboloid** at $\mathbf{x}$. Its first and second fundamental forms agree with the corresponding forms of the surface.

2 The spherical surface $\mathbf{n}(\mathbf{u})$ is called the **spherical image** of $\mathbf{x}(\mathbf{u})$. The squared arc element of $\mathbf{n}$ is called the **third fundamental form** of $\mathbf{x}$,

$$d\sigma^2 = e\,du^2 + 2f\,du\,dv + g\,dv^2,$$

where $e = \mathbf{n}_u^t$, $f = \mathbf{n}_u^t \mathbf{n}_v$, $g = \mathbf{n}_v^t \mathbf{n}_v$.

3 The three fundamental forms and the curvatures K and H are related by the equation

$$K\,ds^2 - 2H \cdot \kappa\cos\varphi + d\sigma^2 = 0 \ .$$

4 The corresponding area elements are related by the equation

$$\sqrt{eg - f^2}\,du\,dv = K\sqrt{EG - F^2}\,du\,dv$$

which establishes a further geometric meaning of K.

5 The torsions of the two asymptotic lines at a point $\mathbf{x}$ of a surface equal $+\sqrt{-K}$ and $-\sqrt{-K}$, see Haack, p.86.

6 An elegant proof of the theorema egregium was given by Baltzer in 1866. Expressing the Gaussian curvature K completely in terms of the derivatives of $\mathbf{x}$, one gets

$$K = \frac{1}{\det{}^2 C}\left(\det\,[\mathbf{x}_{uu}\mathbf{x}_u\mathbf{x}_v]\det\,[\mathbf{x}_{vv}\mathbf{x}_u\mathbf{x}_v] - \det{}^2[\mathbf{x}_{uv}\mathbf{x}_u\mathbf{x}_v]\right)$$

$$= \frac{1}{\det{}^2 C}\left(\det\,([\mathbf{x}_{uu}\mathbf{x}_u\mathbf{x}_v]^{t}[\mathbf{x}_{vv}\mathbf{x}_u\mathbf{x}_v]) - \det\,([\mathbf{x}_{uv}\mathbf{x}_u\mathbf{x}_v]^{t}[\mathbf{x}_{uv}\mathbf{x}_u\mathbf{x}_v])\right) \ .$$

Differentiating the dot products $\mathbf{x}_u^{t}\mathbf{x}_u = E, \ldots$ above gives $\mathbf{x}_{uu}^{t}\mathbf{x}_u = \tfrac{1}{2}E_u$ and further similar expressions. Since

$$\mathbf{x}_{uu}^{t}\mathbf{x}_{vv} - \mathbf{x}_{uv}^{t}\mathbf{x}_{uv} = \frac{1}{2}G_{uu} + F_{uv} - \frac{1}{2}E_{vv} \ ,$$

one has that K depends only on E, F, G and the first and second derivatives of these functions.

7 A net of conjugate curves on a given surface can be constructed as follows: Take any fixed line in space, the tangential cones from the points of the line to the surface, and the pencil of planes through the fixed line. The curves of contact with the family of cones and the intersection with the pencil of planes form a conjugate net.

8 The conjugate net constructed in the previous note for arbitrary surfaces is a generalization of the net of meridians and parallels of a surface of revolution.

9 A surface $\mathbf{x}(u, v) = \mathbf{y}(u) + \mathbf{z}(v)$ is called a **translation surface** since all u-lines and consequently all v-lines are translates of one of another. Since $\mathbf{x}_{uv} \equiv 0$, one has $M \equiv 0$ so that the isolines form a conjugate net.

10 Any one parametric family of spheres envelopes a **canal surface** which is considered in Section 24.6. The circles of contact form one family of curvature lines of this surface.

11 Dupin's cyclides are the only surfaces where both families of curvature lines are circles.

12 Let $\mathbf{x} = \mathbf{x}(u, v, w)$ represent a net of surfaces in space. If this net is orthogonal at every point $\mathbf{u}$, the pairs of surfaces intersect in curvature lines (Dupin 1813).

13 From a projective point of view, a developable is dual to a space curve.

Bibliography

There is an abundance of literature on the topics of this book. The following list is very selective and personal. In particular, it does not imply any conclusion about the references not listed here. Besides the references in the text this list points to other interesting books and articles which either offer further and deeper results or present the material from another point of view.

CLAIRE F. ADLER:
Modern Geometry, McGraw Hill, New York (1967)

KENDALL E. ATKINSON:
An Introduction to Numerical Analysis, John Wiley & Sons, New York (1978)

HENRY F. BAKER:
Principles of Geometry 1-6, University Press, Cambridge (1954)

MARCEL BERGER:
Geometry 1 & 2, Springer, Berlin, (1987)

WILHELM BLASCHKE:
Analytische Geometrie, Wolfenbüttler Verlagsanstalt, Wolfenbüttel (1948) 3rd ed. Birkhäuser, Basel (1954)

WILHELM BLASCHKE:
Projective Geometrie, Wolfenbüttler Verlagsanstalt, Wolfenbüttel (1947), 3rd ed. Birkhäuser, Basel (1954)

WILHELM BLASCHKE & KURT LEICHTWEISS:
Elementare Differentialgeometrie, Springer, Heidelberg (1973)

WOLFGANG BOEHM:
On Cyclides in Geometric Modeling, Computer Aided Geometric Design **7** (1990) 243-255

WOLFGANG BOEHM & GERALD FARIN & JÜRGEN KAHMANN:
A Survey of Curve and Surface Methods in CAGD, Computer Aided Geometric Design 1 (1984), 1-60

WOLFGANG BOEHM & HARTMUT PRAUTZSCH:
Numerical Methods, Vieweg, Wiesbaden, and AK Peters, Wellesley (1993)

EGBERT BRIESKORN & HORST KNÖRRER:
Plane Algebraic Curves, Birkhäuser, Basel (1986)

MANFREDO P. DO CARMO:
Differential Geometry of Curves and Surfaces, Prentice Hall, Englewood Cliffs (1976)

SAMUEL D. CONTE & CARL DE BOOR:
Elementary Numerical Analysis, McGraw Hill, New York (1980)

HAROLD S.M. COXETER:
Introduction to Geometry, John Wiley & Son, New York (1969)

WENDELIN L.F. DEGEN:
Some Remarks on Bézier Curves, Computer Aided Geometric Design 7 (1990), 181-190

NIRA DYN & CHARLES A. MICCHELLI:
Piecewise Polynomial Spaces and Geometric Continuity of Curves, Numerische Mathematik 54, (1988) 319-337

LUDWIG ECKHART:
Four Dimensional Space, Indiana University Press, Bloomington (1968)

GERALD E. FARIN:
Curves and Surfaces for Computer Aided Geometric Design, 3rd edition, Academic Press, Boston (1992)

PETER C. GASSON:
Geometry of Spatial Forms, Ellis Horwood, Chichester (1983) (1989), 313-332

WERNER H. GREUB:
Linear Algebra, 3rd ed. Springer, New York (1967)

WOLFGANG GRÖBNER:
Algebraische Geometrie I & II, Bibliographisches Institut, Mannheim (1970)

PHILIP GRIFFITH & JOSEPH HARRIS:
Principles of Algebraic Geometry, John Wiley & Sons, New York (1978)

HEINRICH W. GUGGENHEIMER:
Differential Geometry, Dover Publications, New York (1971)

WOLFGANG HAACK:
Differentialgeometrie I & II, Wolfenbütteler Verlagsanstalt, Wolfenbüttel (1948), 2nd ed. Birkhäuser, Basel (1954)

J. CARL F. HAASE:
Zur Theorie der ebenen Curven $n$ter Ordnung mit $(n-1)(n-2)/2$ Doppel- und Rückkehrpunkten, Mathematische Annalen II, (1870), 515-548

DONALD HEARN & M. PAULINE BAKER:
Computer Graphics, Prentice Hall, Englewood Cliffs (1983)

CHRISTOPH M. HOFFMANN:
Geometric and Solid Modeling, Morgan Kaufmann, San Mateo (1989)

FRITZ HOHENBERG:
Konstruktive Geometrie in der Technik, Springer, Wien (1961)

JAN J. KOENDERINK:
Solid Shape, The MIT Press, Cambridge (1990) (1960)

FELIX KLEIN:
Vorlesungen über Höhere Geometrie, Springer, Berlin (1926)

FELIX KLEIN:
Vorlesungen über Nicht-Euklidische Geometrie, Springer, Berlin (1928)

WILHELM KLINGENBERG:
A Course in Differential Geometry, Springer, New York (1978)

DAVID HILBERT & S. COHN-VOSSEN:
Geometry and the Imagination, Chelsea Publishing Co., New York (1990)

WELLMANN B. LEIGHTON:
Technical Descriptive Geometry, Mc Graw Hill, New York (1957)

J. CLARK MAXWELL:
On the Cyclide, Quarterly journal of Applied and Pure Mathematics IX, 111–126, (1868)

BRUCE ELWYN MESERVE:
Fundamental Concepts of Geometry, Addison-Wesley Publ. Co., Cambridge (1955)

AUGUST F. MOEBIUS:
Der Barycentrische Calcul, Leipzig (1827), Werke 1, p. 388

JAMES C. MOREHEAD:
A Handbook of Perspective Drawing, Houston Elvesier Press, Houston (1952)

MICHAEL E. MORTENSON:
Geometric Modeling, John Wiley & Sons, New York (1985)

WILLIAM M. NEWMAN & ROBERT F. SPROULL:
Principles of Interactive Computer Graphics, McGraw Hill, New York (1984)

ANTHONY W. NUTBOURNE & RALPH R. MARTIN:
Differential Geometry Applied to Curve and Surface Design, Ellis Horwood, Chichester (1988)

DANIEL PEDOE:
A Course of Geometry, University Press, Cambridge UK (1970)

JOSEPHE PEGNA & FRANZ WOLTER:
Geometric Criteria to Guarantee Curvature Continuity of Blend Surfaces, ASME Transactions, Journal of Mechanical Design (1992)

MICHAEL A. PENNA & RICHARD R. PATTERSON:
Projective Geometry and its Applications to Computer Graphics, Prentice Hall, Englewood Cliffs (1986)

FRITZ REHBOCK:
Darstellende Geometrie, Springer, Berlin (1957)

GEORGE SALMON:
Modern Higher Algebra, G.E. Steckert & Co., New York (1985) (1st edition 1866)

PIERRE SAMUEL:
Projective Geometry, Springer, New York (1988)

HERMANN SCHAAL:
Lineare Algebra und Analytische Geometrie I & II, Vieweg, Wiesbaden (1976)

JORGE STOLFI:
Oriented Projective Geometry, Academic Press, Boston (1991)

GILBERT STRANG:
Linear Algebra and its Applications, Accademic Press, New York (1980)

BU-GING SU & DING-YUAN LIU:
Computational Geometry, Curve and Surface Modeling, Academic Press, Boston (1989)

BARTEL L. VAN DER WAERDEN:
Modern Algebra, Frederick Ungar Publ. Co, New York (1950)

BARTEL L. VAN DER WAERDEN:
Algebra I & II, (in German) 8th ed. Springer, Berlin (1967)

REINHOLD VERO:
Understanding Perspective, van Norstrand, London (1980)

ROBERT J. WALKER:
Algebraic Curves, Dover Publications, New York (1949)

JAMES H. WILKINSON:
The Algebraic Eigenvalue Problem, Clarendon Press, Oxford (1965)

WALTER WUNDERLICH:
Darstellende Geometrie I & II, Bibliographisches Institut, Mannheim (1967)

C. RAY WYLIE:
Introduction to Projective Geometry, Mc Graw-Hill, New York (1970)

Index

A

Index

Index

E

eigenspace 183
eigenvalues 182
eigenvector 183
elevation 35, 269
eliminating the parameter 275
ellipse 115, 190
ellipsoid 141, 250
elliptic point 383
equation of a hyperplane 84, 214
equation of a point 214
Erlangener Programm 79
Euclidean length 158
Euclidean motion 163
Euclidean norm 158
Euclidean space 19, 158
Euclidean vector space 158
Euler angles 44
Euler's identity 306
Euler's theorem 381
exceptional space 225
excluded points 210
exterior of a quadric 132
eye 49
eye distance 50, 61

F

Falk's scheme 11
families of generators 150
fiber of a linear map 17
fiber of a point 225
fiber of an affine map 105
first fundamental form 371
first fundamental quantities 371
first osculant of a curve 318
first osculant of a surface 325
first polar form 307, 311
first polar of a curve 307
first polar of a surface 311
flat point 383
focal conics 198
focal distancees 200
focal line 194
focal lines 191, 193
focal parabolas 199
focal properties of a torus 203
focal ray 201
foci of a quadric 190
focus 192, 194

foot 22
foot point 164
forward substitution 9
four-dimensional space 282
frame 211
Frenet frame 358, 359, 367
Frenet frame continuity 368
Frenet-Serret formula 359
fundamental equations of Gauss and Weingarten 379
fundamental theorem on parallel projections 31

G

gauge quadric 157
Gaussian curvature 383
Gaussian elimination 6
Gaussian normal equations 22, 26
general Frenet-Serret formula 367
general stereographic projection 353
generalized barycentric coordinates 97
generators of a quadric 128
generatrices 386
genus of an algebraic curve 305
geodesic curvature 377
geodesic line 378
geodesic strip 379
geodesic torsion 377
geometric continuity 368
geometric invariants of a curve 367
geometric meaning of extended coordinates 58
geometric meaning of the mapping matrix 15, 16, 59
gradient 278, 279
Gram-Schmidt orthogonalization 160
Gram-Schmidt orthonormalization 160
Greville abscissae 325
G^r-continuity 368
ground line 270

H

Haase's algorithm 245
hard constraints 24
harmonic position 231, 234
harmonic position and polarity 251
Hesse normal form 165
homogeneous coordinates 48, 207
homogeneous of degree 303

Index

Index

Index